# When Coal Was King

# When Coal Was King

## *Ladysmith and the Coal-Mining Industry on Vancouver Island*

John R. Hinde

UBCPress · Vancouver · Toronto

09 08 07 06 05 04 03     5 4 3 2 1

Printed in Canada on acid-free paper

---

**National Library of Canada Cataloguing in Publication**

Hinde, John Roderick, 1964-
When coal was king: Ladysmith and the coal-mining industry on Vancouver Island / John R. Hinde.

Includes bibliographical references and index.
ISBN 0-7748-0935-3 (bound); ISBN 0-7748-0936-1 (pbk.)

1. Coal mines and mining – Social aspects – British Columbia – Ladysmith – History. 2. Coal miners – British Columbia – Ladysmith – History. 3. Strikes and lockouts – Coal mining – British Columbia – Vancouver Island – History. 4. Ladysmith (B.C.) – Social conditions – History. I. Title.

HD9554.C23H45 2003     305.9'622'097112     C2003-905557-4

---

Canada

UBC Press gratefully acknowledges the financial support for our publishing program of the Government of Canada through the Book Publishing Industry Development Program (BPIDP), and of the Canada Council for the Arts, and the British Columbia Arts Council.

This book has been published with the help of a grant from the Canadian Federation for the Humanities and Social Sciences, using funds provided by the Social Sciences and Humanities Research Council of Canada.

UBC Press
The University of British Columbia
2029 West Mall
Vancouver, BC V6T 1Z2
604-822-5959 / Fax: 604-822-6083
E-mail: info@ubcpress.ca
www.ubcpress.ca

# *Contents*

# *Illustrations*

**Maps**

# *Acknowledgments*

It has been a pleasure to write about the history of my hometown, but an even greater pleasure to see the book finally completed. Although researching and writing are solitary endeavours, this book has benefited greatly from the work and advice of others. I am especially grateful to Professor Patricia Roy at the University of Victoria, who welcomed an outsider to the field of British Columbia history and offered me support and encouragement throughout the duration of this project. Both she and Helen Brown at Malaspina University College read the manuscript in its entirety and offered valuable advice and criticism. The work has also been improved through discussions with Patrick Dunae, Gordon Hak, Susan Johnson, and Richard Mackie, all of whom read parts of the manuscript at various times. Both Malaspina University College and the University of Victoria have supported me by offering that most scarce commodity for sessional instructors – employment. Malaspina also provided me with two sessional research grants, for which I am grateful. I would also like to thank Ladysmith artist Michael Dean for painting the illustration on the cover, and Brian Donald at Malaspina for conducting valuable research for me in 2002. Special thanks also go to Jean Wilson and Holly Keller-Brohman at UBC Press, and to Eric Leinberger of the UBC Geography Department for the maps.

This book is dedicated to Kristine and Anthony with love.

# When Coal
# Was King

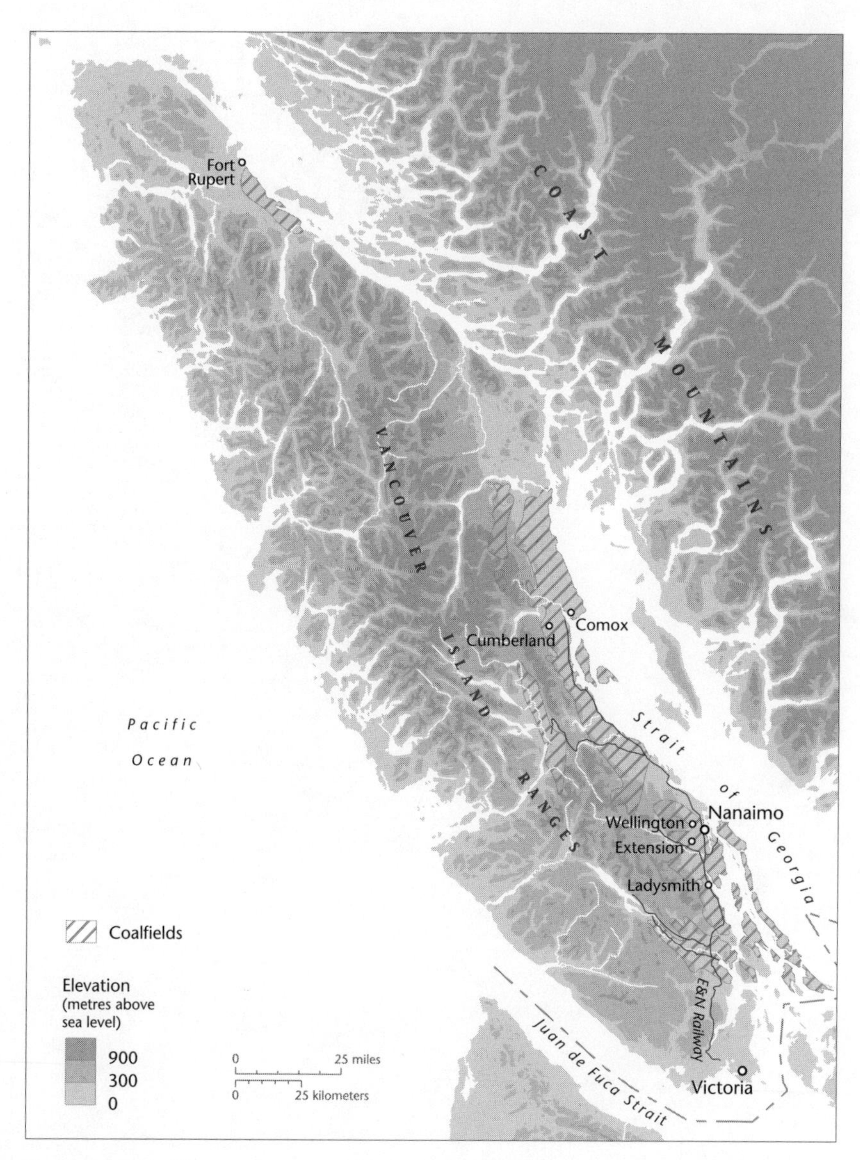


Vancouver Island

# *Introduction*

Coal mining was the most important industry on Vancouver Island during the late nineteenth and early twentieth centuries. From its early beginnings at the isolated settlement of Fort Rupert in 1848, coal production steadily increased, facilitating the colonization of the region by European and Asian immigrants and the industrialization of the Island economy. The coal industry experienced periods of substantial growth in the 1880s and 1900s, with the expansion of the California and domestic markets, although it did not reach its highest levels of production until the 1920s. In 1911, the year before the Great Strike, the Island's collieries employed over 4,600 men and mined a record 1,855,661 gross tons of coal destined for both the domestic and the US market. The three most important mining centres on the Island were the extensive works of the Vancouver Coal Mining and Land Company (VCMLC) at Nanaimo, purchased in late 1902 by the Western Fuel Company of San Francisco, and the Dunsmuir family's operations at Cumberland in the Comox Valley and at Wellington and then at Ladysmith/Extension. In addition, there were a number of smaller mines at East Wellington, South Wellington, Jingle Pot, and later at Cassidy.

This book explores the history of the Island's coal industry by focusing on one community, the town of Ladysmith. Europeans first occupied Ladysmith, or Oyster Harbour, as it was originally called, in 1898. The town site had been purchased two years earlier by James Dunsmuir, the coal and railway magnate, who required a port facility for the coal produced at his recently opened colliery at Extension, located twelve miles to the north. Over the next few years, as the coal at Dunsmuir's Wellington operations was worked out and production began to increase at Extension, he forced his employees to move from Wellington and Extension to the newly surveyed company town that was being built above the wharves and coal bunkers that dominated the shoreline.

Like many other British Columbia resource communities, the town of Ladysmith was the product of "instant urbanization."[1] It was by no means a typical frontier camp town, however. As operations at the Extension mines expanded during the first decade of the twentieth century, Ladysmith prospered and was soon incorporated, becoming an important industrial town and a key transportation hub for Vancouver Island. It was connected to Nanaimo and Victoria by the E&N Railway, which had been completed in 1886, and to Vancouver and the Lower Mainland by the CPR transfer barge. Despite its dependence on the coal industry, the town's economy gradually diversified, and by 1911 it had been transformed from a small company town – in reality a muddy village with more tree stumps lining the steep streets than houses – into a dynamic, ethnically mixed community with a population of approximately 3,295.[2]

The history of Ladysmith during the first two decades of its existence is primarily the story of the coal mines and coal miners. As in coalfields elsewhere, workplace dangers, the struggle for economic survival, conflict between labour and management, and the vicissitudes of community formation shaped the experience of Ladysmith's predominantly working-class population. Independent, tough, and proud, the coal miners of Vancouver Island were reputed to be among the most radical and militant labourers in an extremely polarized province. They formed the backbone of the socialist movement and often engaged in long, bitter strikes. The most violent and protracted labour dispute was the Great Strike of 1912-14, a defining moment for the Island's coal-mining population, which left communities divided and dashed all hopes of union recognition and economic justice.[3]

Despite the meagre gains of organized labour in the province at this time, much of British Columbia labour history has been preoccupied with explaining the alleged radicalism and militancy of the province's industrial working class. According to many accounts, workers on the western Canadian frontier, especially coal miners, accepted "the social relations of capitalism only reluctantly" and were "more inclined to join revolutionary parties and unions than were workers in other regions of the country."[4] This theory of "Western Exceptionalism" has a long history. One of the most influential attempts to explain the uniqueness of western Canadian workers – this particular convergence of militant unionism, radical political activism, and violent strike activity that culminated in the labour revolt of 1919 – was put forward by David Bercuson in his provocative

1977 essay "Labour Radicalism and the Western Industrial Frontier."[5] Over the years, this article has generated much critical discussion. Bercuson based his argument on two disparate theories, neither of which is especially applicable to the coal-mining communities of British Columbia but that have similar objectives, namely, the explanation of a supposed historical peculiarity or anomaly.

The first, the "frontier thesis," was elaborated in the late nineteenth century by the American historian Frederick Jackson Turner.[6] For Turner, the expanding frontier in the American West and the ready availability of free land for settlers, whose existence depended upon "self-reliance, simplicity, and practicality," provided the foundations of the United States' democratic institutions. According to Bercuson, whose vision of the western Canadian industrial frontier is an unambiguous derivative of Turner's thesis, parallel conditions shaped western Canadian history. The frontier was the "great leveller which broke down class distinctions because men were equal, free and far from the traditional bonds and constraints of civilization,"[7] producing an independent spirit in western Canadian resource towns and a mistrust of authority and the capitalist system. The mistrust inherent in the free spirit of the settlers and labourers, who were largely single men from a predominantly British immigrant labour force steeped in a radical heritage, was exacerbated by the brutal conditions of the industrial frontier, which, rather than liberating the individual, offered men only hard, dangerous work with "little upward mobility, little opportunity for improvement. They were not free and were not as good as their masters."[8]

The second theory does not have as long an intellectual pedigree but has greatly influenced coal-mining history. "Isolation theory" was first articulated in the 1950s by Clark Kerr and Abraham Siegal, and holds that certain labourers, such as coal miners, constituted a homogeneous, "isolated mass" that lived "in separate communities, with their own codes, myths, heroes and social standards, and with few 'neutrals' to dilute mass antagonism towards absentee employers."[9] These communities and work camps were isolated geographically from major urban centres and socially and culturally from mainstream bourgeois social norms, values, and institutions, and hence were radical and militant. Although some historians have argued that the physical isolation of coal-mining communities diminished occupational solidarity and reinforced coal miners' natural conservatism,[10] Bercuson argues that the isolation of the miners, when combined with the unique features of the frontier, produced an explosive mixture. Because the "presence of church and other moderating social institutions was weak," and "there were no teachers, clerks, merchants, priests, salesmen, artisans, doctors, or other professionals" on the isolated western Canadian frontier to mediate class conflict, the resource communities were "closed, polarized societies," consisting of a discontented working class on the one hand and a hostile management on the other.[11]

The implications of Western Exceptionalism are far-reaching. It presumes, for instance, that the experience of British Columbia workers was somehow different from that of workers in the rest of Canada and, indeed, elsewhere in the world. While it would be foolhardy to deny the role played by local conditions and events in shaping western Canadian industrialization and the experiences of the working population, it would be equally short-sighted to ignore the similar experiences of workers during this period of profound global economic change. In addition, the theory is based on the assumption that class is not an adequate theoretical category for understanding the peculiarities of the British Columbia resource frontier. Western Exceptionalism seems to have enjoyed currency precisely because it tried to present an alternative to class-based analysis and to divorce economic conflict from the system of economic production, seeking the origins of labour protest instead in external factors. Bercuson, for instance, argues that geography (frontier conditions and isolation), demography (the large proportion of single men employed in the industry), and ethnicity (the radical heritage of British-born coal miners in British Columbia) created conditions conducive to labour unrest. But this reduces workers to the status of passive victims of largely impersonal forces, such as demography or isolation, rather than seeing them as active participants in the daily class struggles that shaped their experience of objective social reality, and effectively displaces class and class struggle, the origins of which are found in the dynamic of capitalist production, as analytical categories and historically determining forces within society. Labour protest is consequently no longer considered within the context of capitalism and capitalist development, but is regarded as anomalous, deviant behaviour. It is a "problem" to be solved rather than a historical phenomenon to be understood, an irrational event rather than a natural feature of modern industrial society, a coherent form of collective behaviour, and a powerful articulation of common aims and interests.[12]

In recent years, scholars have taken exception to this parochial vision of western Canadian labour and working-class history.[13] To cite just one example, John Douglas Belshaw, in a series of seminal articles on the social history of nineteenth-century Nanaimo, has challenged a number of specific arguments of the Western Exceptionalism thesis. Although Belshaw argues that geographical isolation determined the experience of coal miners on Vancouver Island,[14] he has consistently maintained that the conditions of the frontier did not create a natural predisposition towards radicalism or militancy among Island colliers. He has also demonstrated how the apparent "propensity to strike in BC's coal towns" cannot be attributed either to the belief that frontier towns were made up of "single transient workers" with nothing to lose,[15] to the radical and militant heritage of the British immigrant coal miner,[16] or to the poor standard of living of colliers on Vancouver Island.[17]

By exposing these dominant myths, Belshaw and others have obliged the historian to re-evaluate not only the merits of Western Exceptionalism but more importantly the nature of working-class political and economic mobilization on Vancouver Island. In particular, his work ultimately calls for a fundamental reassessment of the experience of class, the workplace, and the transformation of capitalist production. Seen from this perspective, it becomes increasingly evident that the nature of working-class mobilization in western Canada was by no means unique. The advent of industrial trade unions and the growth of socialist and labour political parties during the late nineteenth and early twentieth centuries were part of a universal process of economic development involving fundamental changes in the nature of capitalist production, the function of the state, and in the structure of the working class.

The vital role played by coal miners in this process has been well documented, if not well understood. A glance at strike statistics for Great Britain, for example, demonstrates that while only one-fifth of all strikes were by coal miners, the sheer number of miners ensured that they made up one-half of all strikers.[18] A similar pattern emerges from Canadian strike data. According to a federal government report analyzing strikes between 1901 and 1912 for the Ministry of Labour, of a total of 1,319 recorded strikes, only 100 involved coal miners, or less than 8 percent of the total. Yet the coal-mining industry involved the most employees at 76,572, the next highest figure being from the building trades (67,292), and by far the greatest number of working days lost: out of approximately nine million days lost, over 40 percent, or 3,839,447, were lost in the coal-mining industry.[19] In the Maritimes, a region dominated by the coal industry, miners struck 82 times out of a total of 411 strikes but accounted for 74 percent of striker-days. This has led one authority to conclude that "the coal miners were exceptional not because they decided to go on strike more often than other workers but because their strikes were far larger in terms of numbers and duration."[20]

Ian McKay argues persuasively that the coal miners' position in the vanguard of the labour movement and reputation for radicalism and militancy derived from their central position "within the socio-economic formation." In contrast to other workers, coal miners literally fuelled the industrial revolution. They "were in much the same position as soldiers at the front"; indeed, "they lived on the front lines of the industrial revolution."[21] When coal miners struck, their impact on the economy was greater than that of other workers, not just because of the large numbers involved but because the industrial economy depended upon coal to function. This empowered them more than others, reinforcing their indispensability to a society and economy dependent upon coal for its continued growth and prosperity.[22]

McKay's "power/centrality" theory offers a valuable alternative to Western Exceptionalism. Not only does it reinstate miners as active participants in the social and economic process but it also focuses on the miners' lived experience of economic instability, oppressive employers and management, unsafe working conditions, and unsatisfied expectations, and on their threatened social status and standards of living, and how these factors influenced industrial relations and class conflict. According to McKay, the uniqueness of the coal-mining workplace and the importance of the industry to the economy created a specialized system of production and hence a special work culture that defined the collective outlook or identity of the miners. Their distinctive "mining outlook," based on tradition and precedence and characterized by collectivism and independence, meant that they were more resistant to the threats to their autonomy posed by employers and the state. As McKay argues, colliers were "governed more by historically-sanctioned tradition, precedent, the 'common law' of the pit, than by abstract schemes and formal rules. In both its natural and social aspects, the mine was far more than a mere workplace. It was a dynamic force, even a dynamic tradition, greater than the sum of individuals found within it: a structure dependent upon the integration of its five key systems, but also on the historical precedents its very longevity encouraged."[23] For McKay, periods of economic and social crisis reveal the dynamism of this structure within the coalfields and in society at large, and help explain how it shaped the behaviour of individuals and groups.

While McKay offers a sophisticated and provocative reinterpretation of the history of coal mining in Nova Scotia, he does not completely debunk isolation theory. McKay argues that the collective identity or work culture of the coal miners formed an autonomous culture, a culture that represented a widespread deviation from the mainstream, dominant culture and ideals of the bourgeoisie. However, an autonomous culture (as opposed to a subculture, i.e., a distinct but integrated element of a common culture based on shared meanings) must necessarily exist in isolation from mainstream culture. As a result, McKay risks perpetuating the stereotype of a cohesive community (due to isolation, historical outlook, and occupational concentration) and of an autonomous coal-mining culture distinct from the rest of society. An autonomous coal-mining culture premised on a sharp distinction between proletarian values and bourgeois cultural codes not only minimizes similarities but also smoothes over an array of strategic differences and divisions within the community of miners and the working class itself, such as race, for instance, or skill level, divisions that had to be overcome if the miners were to achieve even a modicum of success on the picket line or in the political arena.

This was rare indeed, however. If the miners shared a common cultural identity of collectivism and independence that was diametrically opposed to the values of the dominant bourgeois culture, then why was this cultural solidarity not readily translated into unified economic and political action? Granted, this relationship is by no means straightforward, as McKay and others have pointed out. There is no reason to presume that solidarity in industrial disputes should be carried over to the political realm. But it does point to the limitations and constraints of labour militancy and radicalism. Clearly labour solidarity was a mile wide but only an inch deep. The common experience of work in the mines was rarely strong enough to overcome the deep-seated divisions within the working class. This partially explains why coal miners on Vancouver Island, while relatively successful in the political field, failed to achieve their goals in the numerous industrial disputes they fought. Success in the realm of party politics clearly did not translate into economic power for the workers. If any one factor was to unite the miners, it was the common experience of work. That it did not reveals the extent to which other factors prevented this collective mining outlook from being translated into effective industrial power.

Without undermining the essential validity of a common work culture based on tradition, precedence, and the structured systems of the workplace, the individual and collective experiences of Vancouver Island coal miners demonstrate that many different subcultures based on ethnicity, gender, work experience, class structures, and class expectations, as well as education and mobility, coexisted within the community. This is not to suggest that the community was cohesive and integrated. There was much room for conflict and argument. Community, after all, requires constant negotiation and dialogue, but dialogue derived from a shared set of cultural meanings or values. Rather than being autonomous, and hence isolated, from the dominant culture, the miners' outlook shared distinct features unique to the form of production but that still functioned within the bounds of the common cultural discourse of society at large.

This book explores some of these issues by looking at the evolution of working-class life and mobilization in Ladysmith during the first decades of the twentieth century. Although Ladysmith was not a typical British Columbia town, there are numerous advantages to studying local history. First, the local context permits a closer understanding of the concrete experiences of Ladysmith's coal miners and their families and reinstates their dignity as active participants in the shaping of their collective destiny. Second, it reveals how local conditions influenced labour relations and the emergence of a clearly articulated political and ideological landscape. Third, studying the coal miners' experience of political, economic, and social struggle helps us understand how they attempted to define and redefine their position in society. The experience of everyday life determined the nature of political and economic mobilization, as well as the aspirations and expectations of the coal miners. Finally, a local study should reveal much about the impact of class, class relations, and class struggle on the lives of the people and on the historical process in general. Class emerges as one of many constitutive and interconnected elements of power relations, including race, ethnicity, and gender; while class was an important social determinant, by itself it cannot entirely explain social reality and the cultural conditions of life.

How class relations shaped the early history of Ladysmith is evident in the town's origins as one of Dunsmuir's company towns. After a brief discussion of Ladysmith's founding and its social structures, I attempt to examine the nature of class, community, and culture in Ladysmith. Despite a common work culture and the existence of a rich associational life, community cohesion was by no means a given. Solidarity did exist during times of crisis, such as the explosion at Extension in 1909 or the strike of 1912-14, but it depended upon the formation of a common identity and set of interests. Confronted with severe economic hardship, a hostile and repressive state, and an inflexible company determined to crush the strikers, miners were able to maintain solidarity only if they had the commitment of the community. However, the enormous pressure of the strikes revealed dangerous, built-in fissures that seriously tested this commitment and eroded community identity and solidarity, preventing the miners from emerging victorious.

Chapters 3 and 4 focus on workplace experience and the problem of mine safety, respectively. To what extent did changes to the structure of the workplace influence class relations and political mobilization? In the past, historians have argued that technological developments, the ethnic division of labour, and the systematic deskilling of the practical miner eroded the miners' traditional authority and autonomy of the workplace, leading to their inevitable proletarianization. Hence, their commitment to labourist or socialist political ideals was a defensive response to these external pressures. Contrary to this view, I suggest that the very persistence, not the erosion, of the "old regime" in the mining industry provided miners with the strength to embrace a political cause and to agitate for improvements in work conditions and wages. The issue of safety, the subject of Chapter 4, is often minimized or ignored altogether when analyzing working-class mobilization, in part due to the belief that the miners were fatalistic and only too willing to sacrifice safety for production. However, this issue was inextricably connected to disputes over wages and was key to mobilizing not only the miners but also their families. As such, it went to the very heart of the mining culture and community.

The final chapters focus on political and economic mobilization and culminate in a lengthy examination of the Great Strike of 1912-14, the longest and most violent strike in the history of the province. Chapter 5 attempts to demonstrate that working-class political culture in British Columbia was refashioned during the first decade of the twentieth century through changes in the economic structure, the development of party government, and the realignment of political discourse along sharp class lines. The growing support of working-class parties heralded the beginning of modern politics in BC, as was the case throughout the industrialized world. A new political culture, based on the clear articulation of group interests, produced new politicians, policies, organizations, and conflicts. Through their experience of political, economic, and social struggle, miners actively sought to redefine their positions in society and their relationship to the dominant economic and political power structures. This became increasingly possible after 1903, when political mobilization and participation accelerated. Over the course of the next decade, a new dynamic evolved that challenged the power of traditional authority and political structures. This political culture was expressed in a new political language that, cloaked in the rhetoric of radicalism, should be seen not just as the articulation of an ideological position but also as an instrument of social and political change.

But did this process produce revolutionaries? This is examined in the last two chapters, which discuss the climax of labour activism in the Vancouver Island coalfields – the Great Strike. This strike was characterized by its violence. Rather than being an episode of unfathomable "revolutionary agitation,"[24] however, the riots of August 1913 reveal the effective limits of the miners' radicalization and militancy. The miners' violence was not an expression of the strikers' power. On the contrary, the striking miners became violent because they had no authority. Viewing the riots as mindless destruction ignores their objectives and meaning and condescends to the past: whether organized or spontaneous, if the riots were mindless, so too were the rioters. Contemporaries were shocked not simply because of the property damage – which of course offended middle-class sensibilities, although they were less concerned about the damage to Chinese property – but because of the message the violence conveyed. Opponents invested the riots with more radical political significance than they warranted because they felt that symbolic inversion could lead to a real inversion of social reality. The strikers, however, never took this step. They specifically targeted certain individuals and groups – strikebreakers, management, and the Chinese – and showed remarkable restraint given the level of frustration and potential for violence, willingly acquiescing to state authority once it was reintroduced into the community and they had made their point. This was no storming of the Bastille.

The eventual collapse of the Great Strike in August 1914 left a legacy of bitterness and hostility in the divided, shaken communities of Vancouver Island; it also led to the demise of the United Mine Workers of America and severely weakened the miners. This was compounded by the First World War, which did little to improve conditions for the miners. Over the next few years, faced with increased competition from oil, the coal-mining industry on Vancouver Island entered a period of crisis and the coalfields slowly declined. Although deindustrialization was by no means permanent – new industries based on forestry eventually replaced the coal mines – the legacy of class conflict had a negative long-term impact on the development of Ladysmith. The strike, followed so quickly by the horrors of the First World War, left an embittered, crippled working-class population. Mobilization stalled, as did the once burning desire to alter the political and economic status quo in British Columbia.

# 1
# *A Selfish Millionaire*

The structures of capitalist economic development on Vancouver Island were well in place before the spectacular economic, social, and political ascendancy of the Dunsmuir family, which played such a crucial role in the region's economy, in the shaping of working-class experience, and in the early history of Ladysmith. As such, the end of Wellington and the subsequent founding of Ladysmith did not mark the beginning of a new era in social, economic, and political relations but rather symbolized the continuation of the conflict between a paternalistic and patriarchal capitalist order and a working-class community struggling to define itself and its interests and values within the global context of late-nineteenth-century Victorian industrial capitalism.

The Dunsmuirs' story has been told numerous times.[1] Robert, the family patriarch, emigrated with his wife to Vancouver Island from Scotland in 1850 as an indentured miner for the Hudson's Bay Company, which had opened its first mine at Fort Rupert two years earlier.[2] After Dunsmuir had laboured for almost twenty years in the Island's mines, his fortunes changed dramatically when he discovered the rich, expansive Wellington coal seam in 1869. In order to capitalize his project, he was forced to form a partnership, which included two Royal Navy officers, Lieutenant Wadham Diggle, who invested the initial $8,000, and Rear Admiral Arthur Farquhar, Commander of the Pacific Fleet, who invested $12,000. In 1873 another navy officer, Captain Egerton, became a partner.

In 1871, with the experienced Dunsmuir at the helm, the firm of Dunsmuir, Diggle and Company began production at the new pits at Wellington, just six miles north of Nanaimo. Over the next decade, the company's operations expanded rapidly. The high-quality bituminous coal was reputed to be as good as the best Welsh coal, and production at the Wellington collieries increased from 16,000 tons in 1873 to 50,000 tons in 1875, most of which initially supplied the ships of the Royal Navy stationed at Esquimalt and those of the Pacific Mail Steamship Company of

San Francisco. In 1879 the company expanded by taking over the neighbouring South Wellington Colliery Company, and by the middle of the next decade, Dunsmuir, along with the rival Vancouver Coal Mining and Land Company (VCMLC), which had purchased the Hudson's Bay holdings in Nanaimo in 1861, had secured a virtual monopoly on coal production on the Island.

The 1880s were years of tremendous growth in the coal industry and success for Robert Dunsmuir and his family. Although Dunsmuir was the dominant partner in the company and had complete control over its daily operations, he did not like sharing profits with his silent partners. By 1879 he and Diggle had purchased the shares of Farquhar and Egerton, and in 1883 Dunsmuir finally bought Diggle out for $600,000, becoming the sole proprietor of the company, which he then renamed Robert Dunsmuir and Sons Company. Shortly afterwards, Dunsmuir embarked upon a course of further expansion. That same year he obtained the coal rights in the Comox Valley and took over the shares in its floundering Union Coal Company,[3] although the full development of the newly acquired coalfields – estimated to contain almost 600 million tons of coal[4] – did not begin until 1888. In 1884 he also opened the smaller Alexandra Mine in South Wellington.

By far the most important and lucrative deal Robert Dunsmuir made during his lifetime was the agreement with the federal government to construct a railway on Vancouver Island. After making a $1,000 contribution or "subscription" to the testimonial fund of Sir Hector Langevin, minister of public works and leader of the Quebec Conservatives,[5] Dunsmuir was ensured a responsive audience in Ottawa for his proposal for an Island railway. In August 1883, after some unexpected delays, Dunsmuir signed a contract with the federal government. In return for building a seventy-eight-mile-long railway from Esquimalt to Nanaimo, Dunsmuir and his four American partners received a $750,000 subsidy, which covered half of the construction costs, and a land grant of nearly two million acres, including all coal and timber rights, worth an estimated $20 million, and a large portion of the Island's remaining arable land.[6] The contract to build the railway and the terms of the land grant were finalized when the Settlement Act was passed by the BC legislature in 1884. On 13 August 1886, Prime Minister Sir John A. Macdonald drove home a golden spike at Shawnigan Lake, completing the Island railroad.

Although the agreement caused outrage in BC and Ottawa, Dunsmuir was unfazed. With the connivance of federal Tories and the aplomb of a veteran card shark, he had brokered the deal of the century. The favour shown to Dunsmuir by the federal government generated enormous personal wealth and enabled him to expand his other business interests, which eventually came to include, besides coal, an ironworks, steamships, tugboats, timber, and real estate. He was also able to consolidate his hold on the Vancouver Island economy even further, leading one middle-class critic to claim that Dunsmuir was "now sapping, as it were, the life-blood of the insular portion of the Province."[7] In an age increasingly critical of monopoly capitalism, he had become the region's "principal monopolist," symbolizing "immoral – as opposed to small-scale and hence acceptable – capitalism."[8] Having come of age, he moved to Victoria in 1883, and four years later began building his ostentatious Scottish Baronial castle, Craigdarroch. He died in 1889, the richest and most powerful man in the province.[9]

Following his death, Robert Dunsmuir's principal interests were taken over by his eldest son, James. By all accounts, James was gruff to the point of being rude; unimaginative, arrogant, and at times recklessly obstinate, he inherited little of his father's business acumen, wit, conviviality, or self-confident manner. For all his faults, his father had been a determined and exceptionally energetic man; his humble origins and capacity for hard work and for even stronger drink earned him at least the grudging respect of his employees. Not so James, whose stint as manager of the Wellington mines from 1876 to 1879, for instance, began and ended with disaster. His refusal to take seriously miners' concerns about the inaccuracy of the weigh scales, upon which their wages and livelihoods depended, led to the first major strike at the Wellington collieries in 1877. Two years later, he ignored the warnings of the provincial inspector of mines, who reported the presence of dangerous levels of gas at the tenth and deepest level of the mine, a mistake that resulted in an explosion on 16 April 1879 that cost the lives of eleven men.

Robert, never sentimental as far as business was concerned, recognized his first son's more obvious shortcomings and replaced James as manager with his son-in-law, John Bryden, a man of considerable mining experience and expertise. Robert also gave more responsibility to his younger son Alexander, who, like his father, shared a passion for hard drink, placing him in charge of the family's important business interests in San

Francisco. To get James out of harm's way, he was, in the words of his recent biographer, "shuffled off to Departure Bay [by his father] where he was given the less demanding job of directing the loading of coal carriers at the company wharves."[10]

To facilitate their business interests, both Robert and James entered provincial politics. During the summer of 1882, while on holiday in Europe and without even mounting a campaign, Robert Dunsmuir was elected to the legislative assembly in Victoria as member for Nanaimo.[11] In 1887 he became president of the council, the most powerful position in the provincial cabinet. In the hallowed halls of the legislature, Robert found himself in the company of like-minded businessmen and other notables, whose main objective, in the understated words of one recent historian, was "the exploitation of regional resources through the provision of public concessions to private interests."[12]

Like his father, James also enjoyed considerable political success, although he professed to dislike the public spotlight and yearned for the life of a landed gentleman, which he eventually achieved. He became member of the legislative assembly (MLA) for the Comox district in 1898, then served as premier (1900-2) and finally as lieutenant governor (1906-9). His stint as premier, which began on 14 June 1900, coincided with a time of political turmoil in BC as the province began the somewhat haphazard transition to formal mass party politics. As the man of the hour, he was expected to usher in a new age of political stability and economic prosperity, but his record as premier was not especially inspirational; his term came to an ignominious end over a railway scandal. Like his predecessors in this office, he had continued the tradition of personal government and was stubbornly incapable of distinguishing between private and public interest. "I have very large interests here," he claimed on one occasion, and "if I don't look after the interests of the country, my own interests will not be looked after."[13]

What James Dunsmuir did not understand was that his interests were not necessarily the same as those of the people over whose lives he exerted considerable influence. He encapsulated his philosophy of life and business in a revealing statement he made to the 1903 Royal Commission on Industrial Disputes. When asked whether he thought his "wealth carried some corresponding obligations with it," the former premier of the province brusquely replied: "No sir. From my standpoint it doesn't."[14] This lack of any sense of social responsibility typified that class of North

American capitalist often referred to as "robber barons" and angered many more progressive individuals, who viewed monopolies as a threat to community values and the social order.[15] For monopolists like the Dunsmuirs, it seemed, the accumulation of wealth and power, by legal and, if necessary, illegal means, was an end in itself.

Not surprisingly, both Dunsmuirs left a legacy of bitter controversy and labour unrest. Their hostility towards organized labour was legendary and set the tone for numerous bitter and protracted conflicts. Neither man was prepared to tolerate the presence of workers' associations or unions at their mines that might in any way interfere with the operation of their companies, reduce their absolute authority, or cut into profits; both were prepared to shut down their works indefinitely rather than give in to the demands of striking miners. The intransigent tone was set early during the 1877 strike at Wellington, when Robert Dunsmuir published his well-known statement in the *Nanaimo Free Press:* "There is an impression in the community that we are obliged to accede to the miners' demands: but for the benefit of those whom it may concern we wish to state publicly that we have no intention to ask any of them to work for us again at any price."[16] For the Dunsmuirs, the question was clear-cut. The capitalist and the speculator, who bore the risks of investment, demanded easy and inexpensive access to resources, government subsidies, and the protection of the state. In return, they expected to exercise complete authority in the workplace and to occupy the dominant position in government and society.

The communities that developed around the grimy pitheads of the Dunsmuir mines were all initially company towns. Wellington and Cumberland (Union) are perhaps the most notorious examples on Vancouver Island, but Ladysmith was also a company town until 1904, when it successfully lobbied Victoria for incorporation. Company towns were common in the coal-mining regions of North America and Europe and usually arose when the mines were located in isolated or sparsely settled areas. Despite the expense of planning and construction, company towns and housing were often necessary to attract labour and ensure production during times of rapid growth. As with other communities, the quality of life in company towns was by no means uniform and was dependent upon a number of variables, such as cost, size, and location. Certainly not all were squalid and vile; some boasted a host of modern amenities, such as schools and churches, had a permanent, stable workforce, and developed into important and prosperous regional centres. Indeed, as one historian has argued, prices in company stores, rents, and quality of housing often had to be competitive with neighbouring towns because of the highly mobile nature of the workforce and the often intense competition for labour.[17] In order to attract workers when labour was scarce, some company towns, especially those close to other urban industrial centres, even offered incentives and other benefits as a means of offsetting an adverse labour market. As a result, company towns often made economic sense to the mine operators, who were keen to begin production, and to the miners themselves, who were not always willing to give up their mobility, independence, and flexibility in the sale of their labour by signing long-term leases or purchasing their own homes.

Such potential economic benefits to the miner and his family were fleeting, however, and came at an important cost. Company towns in the United States and elsewhere had poor reputations.[18] More often than not they were bleak and dismal. With few modern conveniences, workers often lived in wretched conditions. Moreover, workers had few rights and little control over their own lives. Company towns were most notorious because they enabled operators to extend their power beyond the confines of the workplace into the daily lives of their employees. Many unscrupulous owners exercised strict economic control over their workforce through the institution of the company store, and social and political control through company housing, regulations, policing, and sheer brute force.

In Wellington, for instance, the Dunsmuirs owned the land, the works, the miners' homes, and the other businesses. As Bowen has pointed out, Robert Dunsmuir's "approval was needed for every liquor license that was sold and every business that opened and every building that was constructed ... Dunsmuir was the ultimate authority on site and he was determined to control every aspect of the operation right down to where and how people lived."[19] Like most operators, he refused to sell property to his workers.[20] He justified this policy by arguing that the miners would only be wasting their money if they were to buy their homes. Since mining was an inherently unstable business, Dunsmuir "wanted to protect the men from investing money in something that could be rendered worthless if the mines closed."[21] Company housing, however, could be

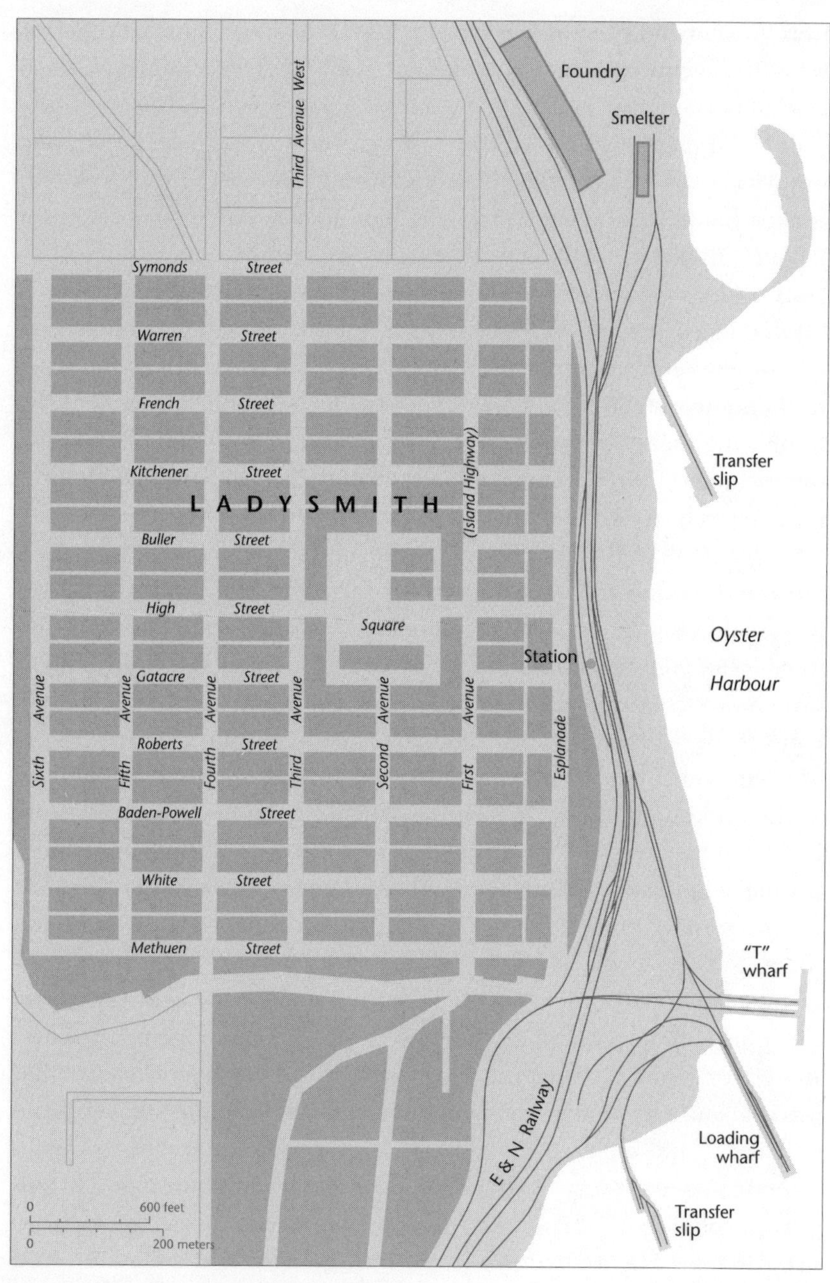


Ladysmith in the early twentieth century

very lucrative, as could the operation of a company store.[22] Dunsmuir even had a gate built across the public road that led to Nanaimo at the point where it met his land, which he closed during the 1877 strike in order to prevent the miners and their families from shopping in nearby Nanaimo, where the price of basic necessities was cheaper, reportedly telling "the butcher not to kill any more beef, the baker not to fetch any bread and the Doctor not to visit us."[23]

One of the more obvious benefits to operators was the fact that company towns effectively prevented collective action because housing was contingent upon employment. Operators found it much easier to replace striking miners and cheaper to house replacement workers since leases could be terminated quickly if the miner was no longer working; this simultaneously eroded the miners' bargaining position with regard to complaints, lockouts, or strikes. With no long-term lease and the threat of immediate eviction always hanging over their heads, miners paid a high price for putting down their tools or engaging in unwelcome union or political activity against the company. Throughout the history of the Dunsmuir mines, union organizers and members were routinely dismissed and strikebreakers hired.[24] More than once, the Dunsmuirs exploited their political influence by calling in the provincial militia to evict unruly workers from company housing and to protect local or imported strikebreakers, as was the case during the strikes of 1877, 1883, 1890-91, and 1912-14.[25] Their control over their workforce extended far beyond collective action. Many, including the Dunsmuirs, also tried to control the political behaviour of the miners. Men often complained that they had been pressured to vote for the Dunsmuirs in provincial elections, and there was considerable fear about what might happen if they did not support the proper candidate. In 1903 the Reverend L.W. Hall, a missionary to the Chinese at Cumberland, claimed that a large number of men had been forced to leave Cumberland because they had voted "contrary to the wishes of the management."[26]

What, then, drew men to Wellington? In part it was a question of wages. In the 1870s and 1880s, at least, Dunsmuir's workers received wages that were competitive with those of the VCMLC, and their housing costs were considerably less. During the 1890s, however, after James took control of his father's interests, wages were reduced, making Wellington a less attractive place to work.[27] Some miners, especially young, single men or inexperienced immigrants, might have preferred the more unencumbered lifestyle in Wellington or could not afford more permanent housing. They could also maintain a degree of flexibility, since they were not tied down by long-term leases or by the burdens of property ownership. Consequently, their mobility, independence, and ability to sell their labour to the highest bidder remained largely intact. This suited Dunsmuir, who initially favoured hiring single, transient men rather than men with families (in 1874 only 8 of 150 men employed at Wellington had families),[28] and unskilled immigrant labourers from various national backgrounds who, like the Chinese, were not readily accepted by the predominantly skilled British colliers of nearby Nanaimo and were less likely to agitate for better working conditions. While this might not have dampened labour unrest among the skilled workforce, it made organization extremely difficult, if only because of the language barrier. As Jeremy Mouat concludes, "such a heterogeneous group would have been difficult to organize, even without the anti-union policies of the proprietors."[29]

Life in the company town might not have been so difficult for the miners to bear had Wellington been an isolated mining camp. But Wellington was situated just a few miles from Nanaimo, where the Island's largest coal deposits were worked by the VCMLC.[30] Unlike Wellington, Nanaimo was not a company town but a permanent settlement of freeholders, with a stronger sense of community, a rapidly expanding infrastructure, and a host of amenities that made life more comfortable.[31] Quality of life and standard of living were generally much better in Nanaimo than in Wellington and Cumberland. Contemporaries noted the bleakness and desolation of Wellington, which consisted of little more than small miners' shacks situated in the dark shadow of the mine works. According to one contemporary, the town was "very wretched and poverty stricken";[32] miners complained about the "hovels built for miners and castles for the Dunsmuirs."[33] Nanaimo, in contrast, blossomed as a city. Whereas Wellington remained very much a primitive coal-mining camp, Nanaimo soon developed the features and amenities of a modern city, with a population of almost 10,000 by the turn of the century. With its electric and gas lighting, it created a lasting positive impression among visitors and residents alike.[34] Many miners owned their own homes, giving them a stake in the community and producing a much more stable, reliable workforce. "Although one of the collieries is practically in the town, the workings extending all under the harbour," one commentator noted, "Nanaimo is the brightest, cleanest, sunniest and most cheerful little place imaginable."[35]

This might have been an exaggeration, but it seemed to capture local sentiment. The relatively benevolent policies of the VCMLC contributed a great deal to Nanaimo's optimism. After the VCMLC bought out the holdings of the Hudson's Bay Company in the early 1860s, coal operations expanded rapidly. Under the eighteen-year-long stewardship of the popular Samuel Robins, who arrived in Nanaimo in the mid-1880s, the company actively pursued policies designed not only to ensure stable industrial relations and the long-term viability of their company but also to foster a strong community. The VCMLC, the anchor of the local economy, created parks, encouraged the development of local cultural life, and sponsored sports. It was more famous, however, for its housing and land-use policies. The firm was the largest landowner in Nanaimo well into the 1890s, and its business included the sale of its land to settlers. Rather than emulate the Dunsmuirs' policy of controlled housing, the land department of the VCMLC had much of its vast landholdings cleared of trees and brush and made five-acre lots available to workmen, usually skilled British colliers, in an attempt to encourage families to settle and cultivate the land.[36] According to one contemporary observer writing in 1892, the "demand for allotments [was] ahead of the survey, which is good evidence of the success and popularity of the land policy of the company."[37]

The land scheme was not without its difficulties. Since the land had to be improved, a number of lots reverted or were sold back to the company, and in 1902 Robins reported that "the sale of lands have been about nil for the last four years."[38] In addition to the allotments, the VCMLC also sold the miners more affordable ⅕-acre lots in the city.[39] The company's land scheme was not intended as a way of making enormous profits and was never used as a weapon against the employees, as Dunsmuir's company shacks were. There were no evictions during strikes, and although the company felt obliged to keep the selling price of the lots on par with city lots, so as not to harm local business interests by forcing the depreciation of the value of city property, by most standards the rent was low.[40] Of course, this policy was not simply a matter of altruism but must be understood within the context of Robins's benevolent paternalism. By fostering the permanent settlement of mining families, the company hoped to create a reliable, complacent workforce. Indeed, as Allen Seager has pointed out, the five-acre lots, which helped offset the contradictions of industrial capitalism, represented a trade-off for capital, as they were designed to "spare the new province of British Columbia from the scourges of industrial poverty, unemployment, strikes, and presumably socialism."[41]

Important as the land scheme was, what ultimately distinguished the Nanaimo operations from the Dunsmuirs' in the 1890s was the VCMLC's radically different approach to labour relations. Simply put, the company based its labour policy on negotiation rather than confrontation and coercion. Prior to Robins's arrival in Nanaimo in 1884, the company struggled to maintain market share and production levels and was plagued by labour unrest.[42] In fact, some speculated that the previous superintendent of the company, John Bryden, had conspired with Dunsmuir to harm the Nanaimo collieries. Following a dispute with the VCMLC owners over the handling of a strike, Bryden, Robert's son-in-law and an ardent admirer of his methods – he wanted to import strikebreakers – resigned his position in Nanaimo only to be hired almost immediately by Dunsmuir as superintendent of the Wellington collieries. A number of contemporaries, including Mark Bate, co-manager at Nanaimo and Bryden's successor, suspected that he "had been working for his father-in-law all along," which may have accounted for the VCMLC's poor economic performance. Either way, many believed that it had been inefficiently managed; it was "a stupid old company" in which the "proprietors had more money than brains."[43]

Robins was determined to transform the fortunes of the VCMLC, and considered improved labour relations to be a key. He took two important steps in this direction. First, in 1891 he recognized the local miners' union, the Miners' and Mine Labourers' Protective Association (MMLPA), signing an unprecedented closed-shop agreement. Granted, the MMLPA was hardly a radical international trade union. But Robins also refused to see the American union, the Western Federation of Miners (WFM), as a threat to local industry, unlike James Dunsmuir and the members of the Royal Commission investigating the strike of 1903.[44] Notwithstanding its political moderation and weakened position following the collapse of the Wellington strike in November 1891, the organization satisfied the current needs of both the workers and management. During the 1890s, when the union was in danger of being disbanded by its members, Robins was instrumental in its survival, much to the disgust of Dunsmuir, who believed that his rival's support of unions had caused wages to rise and had encouraged unrest at his mines in Wellington and Cumberland. Robins, in contrast, recognized that the union was beneficial to the operation of the company, because it facilitated negotiated settlements rather than long and costly confrontations, and created in the process a more reliable and content workforce.[45] And, in fact, the closed-shop agreement with the MMLPA ushered in an era of labour peace in Nanaimo. Not only were the Nanaimo mines free of strikes after the recognition of the union but on at least one occasion the miners even agreed to a wage reduction.[46]

In February 1888 Robins took a second important step to ensure a stable workforce at Nanaimo when he prohibited Chinese labourers from working underground. Robins's exclusionist policy resonated well with the Nanaimo miners and appeared to settle an important grievance. Robert and James Dunsmuir, in sharp contrast, were not exclusionists; they did not abide long by their promise to keep the Chinese from working underground. Cost was their primary concern. The replacement of the Chinese by European workers caused the Dunsmuirs' expenses to rise, cutting into their profits. When the Chinese runners – those who pushed loaded coal cars from the face and brought back empty ones for the hewers – were eventually replaced by Europeans, the subsequent shortage of labour necessitated a wage increase from $2.00 to $2.50 per day.[47] As a result, when Robert Dunsmuir opened his new operations in Cumberland in late 1888, he employed Chinese underground, arguing that the agreement of February 1888 had referred only to his Wellington operations.[48]

The use of cheaper, unskilled labour at Cumberland was a source of great discontent for Nanaimo miners, who continued to agitate against Dunsmuir and the Chinese in the belief that cheap labour, poor working conditions, and longer working hours at Wellington and Cumberland undermined their economic position and the continued viability of their company. Robins acknowledged that Dunsmuir's use of cheap labour had driven his firm out of some markets and given the Wellington collieries an unfair edge.[49] Whether or not this was actually the case remains an open question. The VCMLC was ultimately very successful, and, as one recent historian has pointed out, increasing the size of the Chinese workforce did not necessarily translate into increased per-man production as the operators anticipated would be the case.[50] Despite its alleged tractability and reputation for hard work, unskilled labour, whether Asian or European, could not produce as much coal as the skilled hewers. Any competitive advantage that might have accrued to the Dunsmuirs – and they had many – was probably negated by the more stable workforce and favourable labour relations that existed at Nanaimo.[51]

If, in the words of Allen Seager, "Nanaimo on the eve of the epoch of monopoly capital was, then, a relatively benign environment for the independent collier,"[52] the same could not be said of Wellington and Ladysmith/Extension, where James Dunsmuir exercised a radically different form of paternalism. Dunsmuir's approach to business and to relations with his employees is perhaps best illustrated in the highly controversial events surrounding the closure of the Wellington mines, the beginning of operations at Extension, and the subsequent founding of the town of Ladysmith. Nothing reveals better his lack of social responsibility, his almost complete control over the lives of his workers through the institution of the company town, and his abuse of power and privilege to advance his economic interests.

By the mid-1890s, after more than twenty years of operation, the Wellington works were nearing exhaustion. Fortunately, in 1895 coal had been discovered a few miles south of Nanaimo. The new coalfield was an extension of the vast Wellington seam, and the community that grew up around the new works was subsequently called Wellington-Extension, later shortened to Extension. Dunsmuir's plan was to phase out operations gradually at Wellington and expand production at his new works. Apparently, this was the only part of the plan that was certain. Over the course of the next few years, Dunsmuir's intentions were clouded in confusion and uncertainty, a situation made worse by a lengthy legal battle and by the fact that Dunsmuir played a game of cat and mouse with the local population. Dunsmuir vacillated on two important issues that ultimately dictated the fate of his employees: the shipping point for the new coal and the transportation of miners to and from the new mine. He originally wanted to ship Extension coal from the port of Nanaimo, but this depended upon reaching an agreement with Samuel Robins over the purchase of the Northfield wharves and the adjacent fifteen acres of land. This was not forthcoming, and in early February 1896, Dunsmuir made known his intention to ship Extension coal from his wharves at Departure Bay, the Nanaimo alternative now being too costly. His employees would remain in Wellington and be required to make the short train ride to the Extension mines on a rail line soon to be constructed.[53]

From the perspective of Dunsmuir and the miners, this was certainly the most logical and cost-effective solution. Departure Bay, as Bryden recalled a few years later, was a good, accessible harbour and was well known to customers.[54] Two months later, however, Dunsmuir's plans were scuttled because he once more encountered obstacles with Robins. The problem Dunsmuir now faced was that his proposed rail line from Extension to Departure Bay would have to cross land owned by the VCMLC. In a more hospitable corporate climate, an arrangement between the two rival firms might have been possible. The animosity between Dunsmuir and Robins had not subsided over the years, however, and precluded any sort of settlement. Robins was apparently determined to exact his pound of flesh for an earlier incident from 1888 in which the Dunsmuirs had denied the VCMLC a right-of-way across land owned by the E&N Railway. Now, when James Dunsmuir approached Robins for a similar favour, the manager of the Nanaimo mines refused to grant him permission.

In response, Dunsmuir took the matter to the courts. Pending this decision, he was obliged to steer a different course. In April 1896, he bought 320 acres of land at Oyster Harbour for $6,340 and, in what may have been an attempt to put pressure on Robins, announced his plans to lay out a town site and build a smelter.[55] Five days later, on 1 May 1896, a small but significant change to his plans was made public. There would be no smelter. Instead, the "projected work at Oyster Bay will only include coal bunkers at the present." Dunsmuir, it seems, no longer considered a shipping point at Departure Bay to be a viable option.

This was bad news for the people of Nanaimo. The city council and the Chamber of Commerce were concerned about the impact the loss of the industry and business in Wellington would have on their community. They approached Dunsmuir and asked him to reconsider,[56] which he apparently did, because he confirmed once again that the coal would be shipped from Departure Bay. Plans for a town at Oyster Harbour, he announced, had been abandoned. "Workers will reside in Nanaimo with trains being provided to take them to their daily chores."[57]

In the meantime, work began at Extension. By 1897 two slopes had been driven and prospecting crews discovered good coal between 6 and 15 feet thick; in 1898 work began on the extensive rock tunnel, situated about 100 feet vertically below the mouth of the slope, that would serve as the main entry to the new coalfield.[58] The issue of where the miners would reside was far from settled, however. The game of cat and mouse continued, when on 22 February 1898 it was once more announced that Extension coal would be shipped from Oyster Harbour. Two weeks later, on 11 March, these plans were contradicted and Departure Bay was once more the shipping point. It seemed that the controversy had finally been resolved because just two days later the first piles arrived at Departure Bay for the construction of new wharves. The question was settled once and for all in early October 1898, however, when Mr. Justice Martin granted an injunction against Dunsmuir, preventing the E&N from crossing VCMLC land.[59] With access to Departure Bay now denied, work recommenced on the bunkers and docks at Oyster Harbour and a gang of Chinese labourers began grading a rail line from Extension to the bay. By September 1899 the first Extension coal was being shipped from Oyster Harbour.

The decision to ship Extension coal from Oyster Harbour instead of Departure Bay spelled the end of the old company town of Wellington.[60] Since there was no train to bring the miners to Extension, they were forced to relocate if they wanted to continue working for Dunsmuir. By August 1899 the mines at Wellington were reported to be "practically at a standstill," with only the No. 5 pit operating with a part-time crew. The small miners' houses and cabins stood empty, and the town's few businesses had gone, some to Nanaimo, others dismantled and shipped by train to Extension and Oyster Harbour. By September 1900 the remaining pillars of coal had been removed from No. 5. When work at the Wellington collieries came to an end, the last machinery was removed to Extension and South Wellington. With the pumps gone, the abandoned workings quickly filled up with water. In 1901 the population of the town, which only three years earlier had numbered almost 5,000 residents, had been reduced to a mere 100 people.[61]

As Wellington collapsed, Extension boomed and a small town quickly sprang up in the shadow of the expanding colliery. This did not put an end to Dunsmuir's saga or his cavalier attitude towards his workers. His indecision about the location of a shipping point for Extension coal had caused much frustration, but his decision to force his workers to relocate a second time, this time from Extension to his recently surveyed town site at Oyster Harbour, which in the spring of 1900 was renamed Ladysmith, struck many as pure tyranny.[62] As the *Nanaimo Free Press* put it, the decision represented nothing less than a "tyrannous exercise of authority which had deprived [the miners] of the property in which the savings of years had been invested" and was nothing less than the "denial of the rights of freemen."[63]

James Dunsmuir occupies pride of place in the history of Ladysmith. He is often celebrated as the benevolent and farsighted founder of the town. Streets and buildings bear his name and an elaborate civic myth recalling Ladysmith as "Dunsmuir's Dream" has a firm hold on the popular imagination.[64] But an examination of the events surrounding the founding of Ladysmith suggests that Dunsmuir's actions were neither visionary nor governed by practical economic necessity. On the contrary, his decision to found Ladysmith was largely the result of personal vindictiveness and obstinacy. It was an example of how one man's pride and hunger for control could override economic sense and ride roughshod over the interests and rights of his workers.

By the time work at the Wellington mines had come to a halt, approximately 200 families had resettled in Extension. Others had moved their homes or businesses directly to Ladysmith, which in 1901 had a population of about 750. Dunsmuir later claimed that he had told the miners from the outset that they had to move directly to Ladysmith, but it is not at all clear that an official order was given. According to miner Moses Woodburn, he heard "a common rumour that the men had to go to Ladysmith, but not officially."[65] Indeed, Dunsmuir's dictate became public knowledge in a very haphazard manner. The information that did reach the men and their families was often vague and misleading. Some individuals, for instance, were informed by letter from Dunsmuir after having had the foresight to make the initial inquiry.

Thus, in response to a letter from Thomas McMillan in October 1900, asking permission to relocate the butcher's shop, Dunsmuir replied that he could take the building down provided he removed it to Ladysmith.[66] While this response was indeed categorical, the answer William Gordon received on 1 August 1901 was the opposite. Instead, Dunsmuir suggested that "it is only a matter of time until all the workmen employed at Extension will live at Ladysmith and in view of this I think it would be better for your people not to build there."[67] Similarly, after hearing the rumour that the miners would have to move to Ladysmith, William Joseph and three other men formed a deputation in 1901 to speak with Dunsmuir, who told them that it made no difference where they built their homes, "provided they would suit the officials of the mines. I then spent a few more hundred dollars, on the strength of that, on the house."[68] George

Johnson, a member of the deputation, confirmed this story. According to Johnson, Dunsmuir had claimed that "it made no difference to him" where they lived and that he would not discriminate against them. Johnson even stated that "Mr Dunsmuir told me to build at Extension."[69] Still others heard of his order only by word of mouth or through the foreman. Whichever way the order reached the men, the issue remained uncertain. Not suspecting what was in store for them, many families risked resettling in Extension.

The general confusion was reinforced by Dunsmuir's own contradictory attitude. Initially he did nothing to prevent a number of men from buying property and building homes at Extension. For this there was a precedent. While Robert Dunsmuir had refused to sell lots in Wellington to the miners, James had sold to them, even though he knew that the Wellington mines were almost worked out. Moreover, he did this out of spite. In an embarrassingly unflattering moment, James declared in front of the 1903 Royal Commission: "In my father's time he would not sell any lots [at Wellington], knowing that some day the whole thing would be worked out. At the time of the strike we had ten years ago, the papers took that up. The *Free Press* in Nanaimo said we would not sell town lots, that we wanted to keep the town lots, and it made me angry. I had a survey made and sold so many lots to the miners. Now, as soon as the mines worked out they lose everything."[70]

Over the next few months, after many families had settled in Extension, it became abundantly clear that Dunsmuir was determined to prevent his workforce from residing there. The miners who continued to live at Extension were informed that their services were no longer required. Living in Ladysmith was now a condition of employment. As Dunsmuir brusquely declared, "I made the remark that they could live where they liked, but I could hire them if I liked,"[71] an important qualification that William Jones either did not remember or was not told.

This was a great burden to the miners and their families. Jacob Myllymaki, a Finnish miner, had risked Dunsmuir's wrath and moved from Wellington to Extension, despite receiving a letter refusing him permission to move his house there. He subsequently received a very rocky place in the mines, "an indirect way of telling him to go to Ladysmith."[72] This was by no means an isolated incident. When Louis Astori's place was worked out, he was not given another because he had refused to move to Ladysmith. After working in the mines at Cumberland for two months,

Astori returned to Extension, at which time he was again informed by the pit boss that there was "no need to come and ask for a job if you live at Extension." When Astori objected on the grounds that he did not have the money to move his wife and six children to Ladysmith, the pit boss told him that he would then have to "do what the other fellows were doing," namely, leaving the family in Extension and renting a room in Ladysmith. But Astori was not happy with "batching." He "lived partly up there and partly here." In order to prevent management from learning where his home was, he "circulated through the men on Monday morning, so as to make believe [he was] on the train," even though he had spent Sunday in Extension. This was almost identical to the situation Joseph Fontana found himself in. He was told by James Sharp, the overman of No. 3 mine, that "if on Monday morning I would not be on the Ladysmith train, or at least get off the train, I could not go to work any more." Since he could not afford to quit his job or move his family from Extension, he was eventually obliged to take up lodgings in Ladysmith. "I took a divorce from my wife for a long time," was how he summed up his unfortunate experience.[73]

Some of the miners devised various means to evade the order, such as mingling with the men on the train, but most families had little choice but to relocate, given the enormous pressure they encountered. Still, not all objected to the move. For a few, the forced relocation to Ladysmith was welcomed because the general environment was considered much healthier than at Extension. Ladysmith had a better water supply, was situated at a picturesque location, and was not in the shadow of the mine works. But others did resent Dunsmuir's heavy-handed tactics. Some complained about the distance from their workplace and expressed concern about the pollution from the new copper smelter situated on the waterfront, which apparently made living right next to the pithead at Extension seem healthy. According to Aaron Barnes, the smelter was "a disgrace to humanity"; on some days, depending on the wind, the smoke was so bad "you could cut it with a knife."[74] But the most common objection was the substantial financial burden the relocation entailed. Not only was the cost of living reportedly higher in Ladysmith[75] but the move itself was extremely expensive, especially for those men and women who had invested in property, homes, and businesses. Thomas Isherwood, for instance, had invested his savings in a home at Extension. He subsequently became ill and could not afford to move his wife and five children to

Ladysmith; as a result, he was forced to quit work. Counting on an influx of workers at Extension, William Joseph had invested nearly $1,200 in his boardinghouse. When the community was abandoned, he lost his livelihood and independence. He was eventually forced to sell his house, rent a small home in Ladysmith, and seek work at the mines.[76]

The case of William Joseph may have been exceptional, but most miners faced significant financial stress because they had to pay to dismantle and rebuild their homes in Ladysmith. Dunsmuir did not charge the men for transporting their dismantled homes and personal possessions by rail on the E&N to the new town, but they still had to pay the cost for their re-erection, which ranged anywhere from $150 to $300. Given that the average daily wage for a white miner was between $2.50 and $4.00, this was a considerable sum.[77] This was not all. The miners and their families also needed land on which to rebuild or build their homes. The *Victoria Daily Colonist,* a passionate supporter of Dunsmuir, reported that he had "granted two hundred lots in Ladysmith free to [the miners of Extension] on condition that they build their homes there at once."[78] This was not true. In fact, Dunsmuir had surveyed the town site and divided it into lots, which he then sold or leased to the miners for $100 per 120 by 160 square foot lot. The Wellington Colliery Company also built thirty-five to forty cottages in the new town. The *Daily Colonist* reported that they then "disposed of them to the miners at very reasonable terms: a small monthly payment securing ownership in a few years." The newspaper, overwhelmed by Dunsmuir's largesse and evidently confused about the finer points of political philosophy, declared this to be "a practical application of socialism."[79]

During the testimony of witnesses at the Royal Commission investigating the strike of 1903, it became clear to the commissioners that the forced move to Ladysmith had generated considerable discontent among the mine workers. Although they concluded that it had not been a contributing factor to the strike, they did view it as symptomatic of serious labour relations problems. It was "just such arbitrary and inconsiderate dealing that antagonizes employees to their employers, and is at once a provocation and justification for the formation of unions."[80] Dunsmuir and his brother-in-law, the former Wellington superintendent John Bryden, were called upon to explain their actions. Both men listed a number of reasons for the decision to relocate the company town from Extension to Ladysmith. First and foremost, Dunsmuir claimed, he was acting not in his own interests but in the best interests of his men. Ladysmith was much better suited because the water supply at Extension was poor and unreliable. In addition, the coal seams extended towards Ladysmith, and the works would eventually move in that direction. Finally, by requiring his employees to move from Extension to Ladysmith, he had sought to protect them from union agitators in Nanaimo.[81] Besides the fact that Dunsmuir was indeed trying to secure his best interests, he showed an extraordinarily contemptuous and patronizing attitude towards his employees, who he felt were not capable of independent thought or of determining their own best interests, whether in choosing where to live or with whom to associate. When asked whether "the miner is just as competent to form an opinion as to his place of residence as you are?" he replied, "No, they have to get men to speak for them and judge for them."[82]

These arguments may have satisfied some members of the Royal Commission but they did not get to the heart of the matter. A much more likely explanation for the forced relocation was the fact that Dunsmuir had failed in an earlier bid to purchase the land surrounding the Extension mines, and hence was unable to erect a company town there. This property, amounting to some 200 acres, was owned by Jonathan Bramley, who had bought it from the E&N in 1884. He subsequently sold sixteen lots to miners for $100 each and also leased land for $1 per month to the families who moved from Wellington to Extension. The exact scenario remains somewhat clouded. Bramley claimed that Dunsmuir offered him $5,000, which he rejected. Dunsmuir, in contrast, maintained that he had offered only $2,000, but that Bramley, following the advice of Samuel Robins, proposed to sell it for $10,000, which Dunsmuir refused even to consider.[83] Regardless, Dunsmuir later maintained that he had never intended to build a town there; he wanted to own the land only in order to prevent the miners from living at the pithead.[84] He indicated that it was not a question of cost or land ownership. As he pointedly reminded the Royal Commission, he could have paid the $10,000 for the land and have been no worse off.[85] The fact remains, however, that he did not purchase the land, even though it made more economic sense for his workforce to live at Extension and not to build an entirely new town site.

The final decision was clearly not made out of concern for the welfare of his workers. Dunsmuir saw no problem with the miners living around the pithead at Wellington, although Wellington's close proximity to unionized miners at Nanaimo was a source of aggravation. Likewise, the water supply was not a major obstacle. Engineering problems could have been overcome; according to one witness, funds had already been subscribed for building a water plant at Extension, but Dunsmuir refused to give the provincial engineer final permission,[86] and, as Dunsmuir indicated, cost was not an issue. He had no problem paying $40,000 to supply water to Ladysmith. Rather, in keeping with his personality, Dunsmuir was not going to give in to Bramley, especially after he had rejected Dunsmuir's munificent offer, and allow him to profit by a town site or by leasing land to the miners. The mines, after all, belonged to Dunsmuir. That this might spell the ruin of the presumptuous Bramley could not have been far from Dunsmuir's thoughts.

As Dunsmuir stressed in his testimony, he was determined to control every aspect of his business operations, including where his employees should live. If he could not erect a company town in Extension, then he would erect one in Ladysmith, regardless of cost. He could have owned the land at Extension and built his company town there, but he did not because that would have meant surrendering to Bramley's outrageous demand (a demand apparently made on the advice of his rival, Samuel Robins). His pride prevented him from giving them this satisfaction. Although he stood to recoup some of his expenses through the sale of lots in Ladysmith to his miners – which he also could have done had he bought land at Extension – Dunsmuir was not motivated by costs but by a combination of obstinacy, pride, and power. This was noted, privately at least, by the secretary of the Royal Commission, deputy minister of labour and future prime minister William Lyon Mackenzie King, who described Dunsmuir as "a selfish millionaire [who] has become something of a tyrannical autocrat. To satisfy a prejudice or greed he has undertaken to make serfs of a lot [of] free men by compelling them to live at a place of his own dictation miles away from their work. Owning everything and possessing all but absolute power, he will let his men possess as little as possible."[87]

# 2
## *A Town of Merry Hearts*

The founding of the town of Ladysmith and its rapid initial expansion coincided with a period of sustained national and provincial economic growth that lasted from the late 1890s until just before the onset of war in 1914. The British Columbia economy was fuelled by the increase in the domestic market for the province's natural resources, by the rapid development of the export economy, and by national and provincial policies that actively encouraged immigration and capital investment and suppressed the organization of labour. The most significant developments were in the resource industries, particularly forestry, fisheries, and mining.[1] This economic growth, centred on the Lower Mainland, contributed to a boom mentality based on a vision of endless wealth and economic expansion. More significantly, it marked the beginning of a shift in economic orientation away from the Pacific coast towards continental economic integration with the Confederation, symbolized best by the earlier completion of the Canadian Pacific Railway terminus in Vancouver, the rapid transformation of Vancouver into the dominant metropolis it remains to this day, and the onset of deindustrialization in Victoria.[2]

The expansion of the lumber industry, spurred in large part by the Prairie construction boom and the rapid growth of Vancouver, was nothing less than spectacular. Lumber production, according to one study, increased by 374 percent during the first decade of the century, displacing mining and fishing to become the "king of industries" in the province.[3] Forestry became the largest source of provincial revenue – in 1901 royalties and stumpage fees amounted to $136,000, in 1913 this increased dramatically to $2.25 million – and attracted huge amounts of capital. Whereas in 1901 there were only 32 mills capitalized at $2 million, by 1914 there were 250 mills with capital worth $150 million.[4] The development of the Vancouver Island collieries, in contrast, was not nearly as spectacular and did not match the earlier rates of expansion in the 1880s, when coal production grew from 228,357 tons in 1881 to 1,029,097 tons just one decade later. In 1892, due in large part to the onset of general economic recession, coal production dropped to 826,335 tons. Over the next few years, production continued to fluctuate, but by 1898 output and profits were once again on the rise as the economy came out of recession. That year the Island collieries produced a record amount of coal mined (including sale of coal for coke production) of 1,126,531 tons, valued at almost $3.4 million, a sum that led the minister of mines to boast of a "banner year" for the Island's mines.[5]

Historians who have examined production levels and market conditions have argued that the coal-mining industry on Vancouver Island was "mainly a nineteenth rather than a twentieth century growth industry," and that by the turn of the century it had entered a period of crisis.[6] While the growth rate of the Island coal industry certainly never equalled that of the 1880s or, for that matter, that of the booming forestry sector, it is an exaggeration to claim that coal mining was in a state of crisis at this time. Although oil was certainly making inroads, the coal industry entered a period of crisis only after the First World War. From 1898 until the beginning of the Great Strike in 1912 and the recession of 1913, production and employment levels grew steadily, the exception being the strike year of 1903, which saw total output in the Coast District drop to 860,775 tons. Total output rose from 1,126,531 tons in 1898 to a prewar high of 1,855,661 tons in 1911, an increase of almost 60 percent, and a record number of men, 4,676, were employed at the Island collieries, compared with 2,841 in 1898.[7] In addition, the mines were extremely profitable. Despite the high cost of labour because of its general scarcity, the price of coal was higher. Between 1905 and 1909, for instance, the profits of the Dunsmuir operations amounted to $3,368,845.59, an average of $674,000 per year. These profits were realized in spite of the "generally slack manner in which the business [of the Wellington Colliery Company] has been carried on." In fact, one report suggested that profits could have been much greater had the company opened its own coal depots at the distribution centres: "the sales of the Wellington Colliery Company's coals do not appear to have been pushed and much of the profits ... have gone to the merchants and middlemen."[8]

It has also been suggested that the opening of the Crow's Nest coalfields in the Kootenays in 1898, which eventually almost doubled provincial coal output, resulted in overexpansion during the first decade of the century and, combined with stagnant markets, posed a threat to production at the Island collieries.[9] However, the new mines neither adversely affected the demand for Island coal, which continued to grow during this period, nor disrupted its traditional markets, since the primary markets for Crow's Nest coal were the new railways in the southeastern part of the province and in northern Montana and Washington. Crow's Nest coke went to the new smelters in the Kootenays. In fact, both coal-producing regions were protected from competing with each other because of the high transportation costs from the interior to the coast and, more importantly, because of the higher quality of the Island coal. The Kootenay mines primarily produced run-of-mine or unseparated coal, not the different grades of coal, such as lump, nut, and pea, that were of vital importance in the domestic market.[10] The ability to produce graded coals gave the Island mines a distinct advantage, as the high-quality Wellington lump coal was in great demand in Victoria, Vancouver, San Francisco, and Seattle, where in 1909 it sold for $12 a ton.[11]

Still, the coal-mining industry on Vancouver Island faced significant challenges as its traditional markets declined. Since the early 1850s, California had been the main market for Island coal. In 1903, however, this vital market collapsed. Whereas in 1902, 75 percent of Island coal was exported to California, during the strike year of 1903 only 45 percent of the total production was sold there.[12] It is difficult to determine whether or not this labour dispute had an impact on the collapse of the California market. While the mines at Extension remained closed from 12 March to 3 July 1903, production continued at Union and Nanaimo.[13] More to the point, sales to California had been declining before the strike. In 1900 the Island collieries shipped a total of 766,917 tons to California; in 1902 this had dropped to 591,732 tons, and in 1903 the total was 289,890. This trend continued over the course of the decade. In 1911, a record production year, only 21.6 percent of Island coal was exported to California.

The main reason for the collapse of the California market was the general decline in coal consumption in the state from a total of 1,889,128 tons in 1900 to 1,215,554 in 1903. This was due to the increased use of fuel oil, especially on the railroads, and to the high price of coal, both of which were a cause for concern to Island residents.[14] Ultimately, the impact of the collapse of the California market on the Island collieries was only short term. Sales quickly rebounded due to the substantial increase in domestic consumption brought about by the expansion of the railways in British Columbia, in particular the logging railways, and the rapid urbanization of the Lower Mainland. By 1911 the domestic economy consumed over 76 percent of the Island coal, which more than made up for the loss of the American market and actually enabled production to increase.[15] Indeed, rather than overproducing for the needs of the province, the Island collieries were sometimes unable to satisfy the high demand because of a shortage of skilled and unskilled labour. In 1906 the minister of mines reported that, riding the "wave of general prosperity which has swept over the country," British Columbia collieries "find themselves in such a position that they have more orders than they can fill." Officials in Victoria feared that a "fuel famine" was "imminent," and in 1907, 3,000 tons of coke had to be imported from Australia.[16] Likewise, in 1910 provincial mineralogist William F. Robertson reported that the "market is growing faster than the collieries are being developed." One consequence of this growing demand was the high price of coal. This was a decidedly mixed blessing. The "maintenance of the present high price of coal on the seaboard," Robertson wrote, was "a consoling thought to the mine-owners, if not to the consumer." In fact, he somewhat awkwardly explained, the high demand had "permitted of the domestic price being kept at a figure so high as to admit of the importation from California of fuel-oil as a competitive fuel."[17] Consequently, while growth rates may not have reached the spectacular level of the 1880s and production fluctuated between 1903 and 1906 due to the strike and market adjustments, overall the long-term economic prospects for the coal-mining communities on Vancouver Island during the first decade of the century were very sound.[18]

This favourable economic climate gave rise to considerable optimism in the Island's resource communities. Nowhere was this more the case than in the new town of Ladysmith. Visitors and journalists alike expressed considerable enthusiasm about the young community. The growing town, one booster observed in early 1901, was "hustling and bustling," a potential "rival to the cities of Victoria and Nanaimo, a town with a brilliant future."[19] A few months later, the same newspaper, convinced of the town's prospects, enamoured by its scenic beauty, and impressed by the "well dressed women and healthy, robust children," concluded that Ladysmith was "a town of merry hearts, which is bound to go ahead, for one hears laughter, song and whistling everywhere."[20]

Despite the town's rapid expansion, its dependence upon the coal industry, and the numerous taverns and hotels that lined its dusty streets, writers also took pains to point out that Ladysmith was not a stereotypical "boom" town or squalid, rough-and-tumble camp of the western Canadian industrial "frontier." "Unlike the majority of new mining towns," the *Daily Colonist* reported with a great deal of reassurance and no doubt considerable relief, "Ladysmith is an orderly, law abiding community" with a constable whose "services are seldom or never required."[21] This was not how the local residents who were lobbying for police services felt. Ladysmith was not as law-abiding as the journalist from Victoria believed. The chief constable at Nanaimo certainly did not think so, writing in October 1900 that "as there are some two or three hundred people here [in Ladysmith] now, and quite a few drunks around the streets by times, and sometimes quite noisy, also some few tramps that wants [sic] a watchful care over them ... [w]e think that a Policeman would be of great service to Ladysmith."[22] A few years later, the policing situation was once again the subject of great concern, because in June 1905 the community was once more without a constable. The constable had been charged on suspicion of stealing $31 from a drunk. He was tried, dismissed, rearrested, tried again, and then set out on bail – hence the vacancy. The town council evidently thought a conspiracy was afoot. A number of people were suspected of the theft, but everyone seemed to be lining up against the hapless police officer. As a consequence, council informed the police commissioner that the town would be without a policeman for the rest of the year and that his salary would be spent on the town's streets.[23]

If the lack of a constable meant that the streets of Ladysmith were no longer quite as safe as before, at least now they could be properly sidewalked and graded, a sure sign of progress. The local media celebrated the fast pace of development and continually emphasized the town's permanence and potential. Reporters commented with evident pride on the "substantial nature of the buildings," many of which, like the Anglican church, had been dismantled and transported from Wellington and Extension and then reassembled in Ladysmith, and praised the ambitious and industrious nature of the town's businessmen. For many of its citizens, the arrival of banks, real estate offices, salesmen from Victoria, and other services was a clear sign that Ladysmith had come of age; their presence was a vote of confidence in the town's future. The existence of a branch of the Canadian Bank of Commerce was for Ladysmith, as it undoubtedly was for many other small communities, "evidence of the faith of a powerful financial corporation in the stability and permanency of the town."[24]

This optimism appeared to be justified by the expansion of Ladysmith's industrial base and population. Within a short time, a host of new industries was established on the waterfront, including a sawmill, the Tyee Copper smelter, and an ironworks and foundry. The most important of these new industries was logging, which soon developed into the second largest employer in the region. A number of logging companies, the largest of which was the Victoria Lumber and Manufacturing Company, harvested trees in the hills to the west of the town, providing employment for almost 200 men in 1910. The logs were then shipped by rail or boom to the company's sawmill in Chemainus.[25] Ladysmith even had a brewery, an opera house, and its own small cigar factory, opened in 1908 by Percy K. Winch, which employed "a large number of hands" and made the "Grand Duke," a "Havana that delights all smokers from the boy behind the shed flirting for the first time with Lady Nicotine, to the smoker of half a century."[26]

But it was the Extension colliery, thought to be "almost endless in extent," and the "unceasing" demand for coal, that was the main source of optimism. Neither occasional labour unrest, nor market fluctuations, nor even the sale of Dunsmuir's holdings in 1910 to the eastern Canadian railway tycoons William Mackenzie and Donald Mann for nearly $11 million – seen by some recent historians as a prescient sign that Dunsmuir was anxious to get out of an industry that was in obvious decline[27] – could dampen local spirit. Instead, it held out the prospect of even greater investment, employment, and industrial expansion. As the *Ladysmith Chronicle* commented, it promised nothing less than "a new era of progress and prosperity."[28]

According to one American travel writer, the new town of Ladysmith was a picture of class harmony rather than class conflict. "The miner here mingles with the capitalists and the poor are upon a par with the rich," he wrote in the spring of 1907. In Ladysmith there was "little distinction between one another."[29] Despite this rosy vision, Ladysmith was not a town of merry hearts, a town in which the poor lived on a par with the rich and where there was little distinction between the two. As in most communities in

The coal-loading facilities at Ladysmith, early 1900s.
*BC Archives, E-04891*

the making, conflict and consensus existed side by side in Ladysmith as individuals and groups sought to negotiate and articulate their identities within the context of a modern industrial, urban society. Miners and their families predominated, but we must avoid the temptation to stereotype Ladysmith as a typical coal-mining town. The culture and community that emerged were characterized not just by class division but by ethnicity, race, religion, and gender. Even here, however, the lines were never clearly drawn. Rather than being a community polarized (and by implication paralyzed) by division and tension, the defining feature of Ladysmith was instead the dynamism and fluidity of its social relations.

The pattern of social relations that emerged in Ladysmith during the first decade of its existence was similar to that of other towns and cities in Canada at the turn of the century. Although the vast majority of the town's population consisted of wage earners employed at the Extension collieries, a small but highly visible middle class of merchants, tradespeople, and professionals such as doctors, teachers, lawyers, and clergymen, played a disproportionate role in local society. This small group of notables tried to define power relations and reinforce its authority and status through practices that tended to exclude individuals from "respectable" society on the basis of occupation, ethnicity, and gender. As elsewhere, the predominantly British middle classes conflated their interests with those of the community as a whole; they presumed to be its natural leaders and the guardians of its values and virtues. The town's newspaper aggressively promoted local commerce, celebrated the "vim and vigour" of the local shopkeepers, and, in keeping with the progressive image of the town, was impressed by how the merchants kept abreast of the latest "business methods." More to the point, the paper reminded its readers that the shopkeepers were advancing "not only their own interests, but also those of the community."[30]

In many respects, the social elite of the new town was ready-made. Coal miners and their families were not the only people transplanted to Ladysmith when the mines at Wellington were abandoned. A number of small tradesmen, merchants, and hotel owners had also dismantled their businesses and transported them on flatbed railcars to Ladysmith. Once they had set up shop again, they began to redefine their social, political, and economic space, energetically promoting the town and their commercial interests. Development was quite rapid between 1899 and 1904, and the middle class played a leading role in creating the town's infrastructure. A commercial district was soon established along First Avenue, and schools and churches of the four major denominations – Anglican, Catholic, Presbyterian, and Methodist – were built or rebuilt on land donated by Dunsmuir. In addition, important services, including water and telephone systems, were soon in operation.[31]

Middle-class political pre-eminence quickly followed. One of the most important vehicles in this process was the board of trade, which was first

First Avenue, Ladysmith's commercial district, looking north, at the turn of the century.
*Photo courtesy of Ray Knight*

organized in 1902 by a committee of prominent property owners. In addition to promoting the town of Ladysmith as a place to do business and even as a tourist destination, the board of trade energetically campaigned for incorporation, which it finally achieved in June 1904. Headed by John Coburn, Ladysmith's first mayor, owner of the Ladysmith Lumber Company and former conductor on Dunsmuir's E&N Railway, the board of trade and the movement for incorporation sought to establish the town's independence from Dunsmuir's control and influence. Dunsmuir, who at this time was divesting himself of his industrial interests on the Island (in 1905 he sold the E&N to the Canadian Pacific Railway for $2.3 million) put up little opposition to the plans of the board of trade. His primary concern was that his extensive industrial holdings not be included in the municipal boundaries, a concern that was shared by Coburn, who also managed to get his lumber mill on the waterfront excluded.[32] The main source of resistance to incorporation came from the miners, who were legitimately concerned that they would have to bear the additional

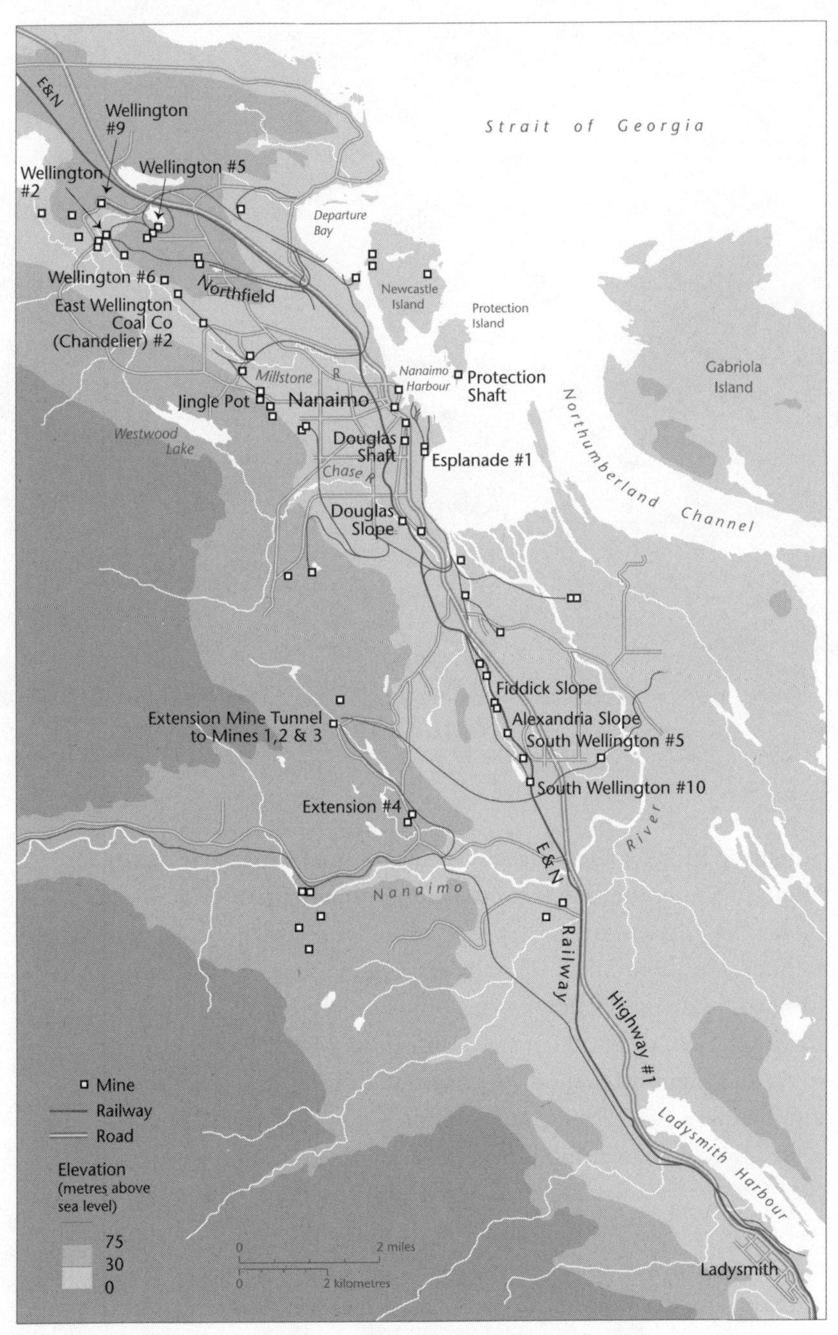


Main coal mines of central Vancouver Island

costs. The exclusion of the main industries from the municipality reduced the town's tax base substantially and shifted the burden of taxation squarely onto the town's working-class population. Only after a lengthy campaign and four petitions did the board obtain the number of signatures it required.[33]

The success of the merchants and professionals in the drive for incorporation quickly translated into local political dominance. The principal agitator for incorporation, John Coburn, was chosen mayor of the town by acclamation, although his business did not contribute to the town's tax revenue. The majority of the town's population was at a distinct disadvantage. The workers not only bore the brunt of taxation but were also excluded from political power because of a set of undemocratic property qualifications that restricted their participation in local political life. Very few miners and labourers could afford the $500 of insured real estate required for aldermanic candidacy, and even fewer the $1,000 required for mayor; as a result, local government was, with the odd exception, dominated by the town's middle-class elite.[34]

In his study of the resource towns of Port Alberni and Prince George, historian Gordon Hak concludes that "these were tightly knit environments and the respectable citizens could ignore the resource workers, the ethnic outsiders, and the unskilled."[35] The attempt to replicate a middle-class milieu characterized by social separation can be seen not just in the evolution of local political life but in the residential landscape and the makeup of community associations. Members of the middle class, for instance, tended to live in more comfortable houses on larger, more expensive lots east of Second Avenue towards the Esplanade, much closer to all amenities. Most miners and their families, in contrast, lived in small houses on the hill to the west of the business district, on lots sold to them by Dunsmuir. The physical separation was reinforced by social and cultural divergence, as seen by the number of middle-class organizations, such as the board of trade, membership on church committees, or even some types of sporting and social events, which were the preserve of the local elite. For instance, whereas the congregation of the Methodist and Presbyterian churches consisted primarily of British miners and the congregation of the Catholic Church of Eastern Europeans, the congregation of the Anglican Church consisted mostly of the merchants and professionals of English extraction. It offered a host of suitable activities for "respectable" citizens, ranging from choir evenings to strawberry and

The Esplanade, now the Trans-Canada Highway, looking north, at the turn of the century.
*Photo courtesy of Ray Knight*

ice cream socials, picnics, and excursions aboard the CPR steamer *Joan.* While the miners enjoyed success on the soccer pitch, the "Church of England Tennis Club" was remembered as being "always in demand by the young businessmen"; among the prizewinners in 1905 were the bank manager, W.A. Cornwall, the daughter of Ladysmith's mayor, and the Reverend Richard Bowen.[36]

The middle class occupied a pre-eminent position in local society, but workers were not excluded from community life or respectability. Class structures were quite fluid, especially for the skilled European colliers and their families. As part of an elite "labour aristocracy," they occupied a somewhat nebulous position between the working class and the lower middle class of craftsmen and small businessmen. Colliers or practical miners were often subcontractors of labour, owned their own homes, and were deferred to by management and the unskilled because of their experience, skill, and knowledge of the work process. They viewed themselves as producers who through honest but hard work could advance themselves and their families, and were considered to be the natural leaders of the mining population and its organizations and associations. As one historian has pointed out, miners on Vancouver Island had "expectations of self-improvement" and upward social mobility. Many took advantage of opportunities for advancement and left mining altogether in order to run their own businesses, such as pubs and boardinghouses, or engaged in other forms of trade and commerce.[37]

These attitudes reinforced the miners' more conservative, labourist impulses. Labourism represented a distinct working-class ideology that allowed skilled workers to view themselves as a vital component of the existing capitalist system; their political objectives were not to destroy capitalism but to gain a more equitable share of the capitalist pie. Moreover, these attitudes indicate that a miner's identity was not simply shaped by class. Status, expectations of upward social mobility, and in many instances ethnicity not only differentiated miners from the unskilled labourers below them; they also created a bridge to the middle class, whose values and culture they partly shared, and enabled them to participate on their own terms in bourgeois society at large. As a consequence, miners viewed themselves not just as members of the working class but as members of the community, as citizens. This was reflected in Ladysmith – in sharp contrast to Wellington and regardless of the political clout of the business elite – in the emergence of a strong, inclusive civic culture that fostered cross-class relations. This civic culture manifested itself in a variety of ways, including the many lodges and friendly societies, the intense pride of place, the optimism in the future of the community, and the rivalry that developed between Ladysmith and Nanaimo.[38] It tended to reflect, rather than refract, the dominant middle-class values and ideals and to transcend class divisions. The emergence of a strong civic ideology, combined with the conservative, labourist impulses of the status-conscious skilled workforce, may have prevented the emergence of a true oppositional subculture in Ladysmith.

Other factors contributed to the development of this civic culture. The sheer number of workers and their families, for instance, demanded inclusion, even if this was not without conflict. Through their sports and hunting clubs, their participation in the numerous lodges, friendly associations, and social clubs such as the town band and choir, their membership on local committees, and their limited participation in municipal politics, the miners contributed actively to the making of the community.

Their economic strength (after all they were the merchants' main customers) also necessitated inclusion; it is no surprise that local merchants and producers, repelled by the power of monopolists like the Dunsmuirs, who restricted economic opportunities, often sympathized with the miners' struggles for economic justice.[39] In small, close-knit communities like Ladysmith, social status, for Europeans at least, was determined less by a priori conceptions, such as family or economic position, than by standing derived from public service and commitment to the community. Through their participation in and contribution to community life, miners and their families were able to break down class barriers and preconceptions about status and achieve a high degree of social respectability.

The social landscape of Ladysmith was also shaped by a remarkable ethnic diversity. Ethnicity was an important element in the formation of working-class identity. Ladysmith's population was notably heterogeneous, characterized less by the racial polarization that was the primary feature of early Wellington and Nanaimo than by a new multi-ethnicity. The majority of the town's residents – more than 60 percent – were British in origin, including most of the local professionals and businessmen. The 1911 census enumerated 981 Scots, 891 English, 199 Irish, and 103 "other" British residents. The dominance of British miners has been noted by other historians and was typical of many coal-mining communities in North America.[40] However, if miners of British origin traditionally predominated in these mines, the coal-mining industry also attracted other immigrants, and as the population grew, the proportion of British-born population shrank.[41] Approximately one-third of Ladysmith's population at this time consisted of other Europeans, the majority of whom also worked in the mines. The largest ethnic group after the English and the Scots came from Russia (304 residents), although most of these were from Finland, which until 1917 was part of the Russian Empire. There were also 209 people enumerated from the Austro-Hungarian Empire, the vast majority of whom were Croatians, followed by 156 Belgians, 109 Italians, and smaller numbers of Germans, French, and Scandinavians.[42] There was also a small group of African Americans and a much more substantial number of First Nations living in the district. African Americans, many of whom had come to the region during the Cariboo gold rush of 1858, worked as unskilled labourers in and around the mines and wharves, although some established their own businesses. The best known was J.X. Smith, who owned a restaurant and a boardinghouse. Smith gained notoriety in the town because he was suspected of having an illegal still and of operating a brothel out of his boardinghouse, which was often raided by the provincial police.[43]

The First Nations population lived in permanent settlements at Kulleet Bay and Shell Beach on the eastern shore of Oyster Harbour. During the nineteenth century, according to some European accounts and oral histories, the harbour was the scene of a number of intertribal conflicts.[44] The harbour and area of the future town site were also important hunting and fishing grounds, and Long Island, across from Shell Beach, served as a burial site: "here corpses were slung in trees with their guns, axes, spears and all the implements of the hunt on the ground below."[45] With the arrival of Europeans in the region, the First Nations population experienced significant dislocation as settlement and industrial activity disrupted its traditional economic and social patterns. Once the vibrant site of feasts and potlatches, Shell Beach, part of the Indian Reserve, was in a state of crisis. The tribal meeting hall, the "Rancherie," had fallen into "picturesque ruin." According to a revealing article in the Ladysmith newspaper from late 1910, which neatly captured the prevailing colonial attitudes, the "Rancherie," "together with the small Indian graveyard, with its rudely carved crosses and tombstones form[ed] a sharp contrast with the smart houses and business blocks of the enterprising city on the opposite side of the bay. They are the connecting links between the fast dying race of aborigines and the increasing population, the future possessors of the land."[46]

Despite serious economic, social, and cultural decline, First Nations peoples were not made "irrelevant" by European conquest and settlement, as some have suggested, although they were increasingly marginalized.[47] The First Nations population was generally considered unreliable as a source of labour by Europeans because they did not adhere to the more rigorous – and exploitative – demands of the capitalist work ethic. Instead, they chose to work according to their own sociocultural rhythms, moving seasonally to the fishing grounds of the Fraser River. The available evidence suggests that despite racial discrimination and other social pressures, such as the introduction of alcohol, local First Nations were able to adapt their lifestyles to the demands of the capitalist economy while preserving their cultural integrity and sense of community.

This took a number of forms. During the summer, some First Nations peoples engaged in small-scale entrepreneurial activity, such as fishing, digging clams, or harvesting seaweed from the harbour and environs, which they then sold to the Chinese communities in Victoria and Vancouver. The money that was earned, according to one source, was then pooled, and basic items such as tea, sugar, and other winter supplies were bought in Ladysmith.[48] Others sold their labour for wages or trade goods. As numerous authors have pointed out, First Nations labour was "essential to the capitalist development of British Columbia," especially during the colonial period.[49] First Nations men and women were employed mining coal at Fort Rupert and Nanaimo, and the women "were prized as mates for white men in a bride-poor frontier town," although this did not enhance their status in the eyes of the white population.[50] First Nations at that time constituted an estimated 5 percent of the workforce in the mines, performing mostly unskilled labour, such as hauling and loading above ground or at the docks.[51] During the late nineteenth and early twentieth centuries, local First Nations peoples continued to work on a casual basis for the mining companies and were hired by Dunsmuir when he was constructing the railway and the wharves at Ladysmith. The experience of Wawa Wawa Latza, or "Big Joe," does not seem to have been atypical. Born around 1872, in addition to a host of traditional activities, such as hunting, fishing, and canoe building, he worked laying track for the railway and then as a carpenter in Westholme, where he apparently earned the incredibly low wage of fifty cents per day building houses and boats.[52]

Most of the European minorities also tended to be isolated socially, economically, and physically from "respectable" British society, which often discriminated against them, although by no means to the same degree as the First Nations population they supplanted.[53] They often formed distinct cultural groupings within the working-class population. The Finns, many of whom had worked in Wellington and other industrial communities in North America before settling in Ladysmith, lived in their own enclave known as Finn Town. They had their own community hall for meetings and social events, and actively sought to preserve their cultural identity through Finnish language classes, folk music and dancing, and a host of other events.[54] As with the Finns, the substantial number of Croatians formed a distinct subculture within the mining population. They also tended to live in the same neighbourhood and belong to the same social organizations, and actively attempted to preserve their cultural heritage and identity. One of the most important vehicles in this process was the St. Nicholas Lodge of the National Croatian Society, formed in

Ladysmith in October 1903. Besides hosting many cultural, educational, and social activities, the best remembered of which was the very popular "tamburitza" orchestra, the lodge served as a benevolent society, providing its members with accident and medical insurance.[55]

Although the various nationalities formed vibrant subcultures, there was still considerable social, cultural, and economic interaction between the groups. Foreigners were not automatically excluded from local society. Special occasions and celebrations, such as Christmas, provided opportunities for residents to forge broader community ties and break down ethnic barriers, as did participation in the churches, the numerous friendly societies, and other social gatherings. The most important factor uniting the various European peoples, however, was the common experience of work in the mines. The dangers of the mine and the dependent nature of the work meant that mutual cooperation and understanding were at a premium. This solidarity transcended ethnic divisions and was often carried over to the struggle against the coal companies for economic justice. British miners welcomed immigrant workers, with the exception of the Chinese and Japanese, into their organizations, often providing interpreters at union meetings and during strikes, and miners from the various nationalities were very active in union organization. During the 1891 Wellington strike, the committee of men that went to interview the mine superintendent, John Bryden, was composed of "one German, one Italian, one Belgian, Carter, and one Russian Finn," or "one from every nationality that was working here at the time."[56] Years later at Ladysmith, Bill Keserich, a well-respected Croatian miner and founder of the St. Nicholas Lodge, who lost his life in the explosion of 1909, was an active union member and sat on the executive of the local branch of the Western Federation of Miners.[57] Likewise, the various nationalities played a major role in the Great Strike of 1912. Much of the alleged radicalism and the disturbances during the strike were attributed to the presence of the various non-British ethnic groups in the district.[58]

The principal target of racism in the mining communities of BC was the Chinese, the one group that was excluded from European society. Ladysmith was home to approximately 177 Chinese residents, most of whom worked as unskilled mine labourers at Extension or were employed on the wharves and in the mills and logging camps. As in other communities, the Chinese were generally unwelcome and isolated. They occupied the lowest position in the town's social hierarchy, and literally lived on the very fringe of society. In Ladysmith's segregated urban landscape, the

Chinese lived first in bunkhouses on Methuen Street and later, when they had to move to make room for a park, outside the city limits. This isolation, forced as it was, accentuated the vast cultural and social gulf that existed between the Europeans and the Chinese, but also ensured that the small Chinese community became relatively self-contained and self-reliant. Unwilling to be integrated into a society that despised and exploited them, the Chinese, perhaps more than any other ethnic group in the mine towns, preserved their culture and the bonds of mutual dependence.

Most of the Chinese residents of Ladysmith were single or had come to Canada without their wives and children. Within this group, however, there was also a small number of more prominent, prosperous Chinese businessmen. Unlike the majority of the Chinese labouring population, this tiny merchant middle class had families in Ladysmith and formed a social and economic elite within the Chinese community. The six or seven Chinese families and businessmen played a vital role in this community, facilitating interaction between the Chinese labourers and society at large, as, for instance, when the merchant and grocer Mar Sam Sing hired an accountant to help Chinese labourers with their time sheets, or providing social assistance and relief to those in need.[59]

The situation of the Chinese residents was unenviable. Besides being vigorously discriminated against and exploited by the state and by coal-mining companies that employed them at half the wages of the white miners, the Chinese were often subjected to racist attacks in the course of their daily lives. While anecdotal evidence suggests that there was some degree of interaction between the Europeans and the Chinese – for instance, Europeans often expressed a curiosity about certain "exotic" customs, such as Chinese New Year, and a much keener interest in opium and gambling dens[60] – other evidence suggests the pervasiveness of racial prejudice. Indeed, intolerance seemed to be deeply ingrained, even in the attitudes of the children. One account, chronicled by Viola Johnson-Cull, reveals the mix of interaction and learned behaviour: "I think our sole pleasure on Saturdays and holidays was trying to teach the Chinese boys how to play football. Perhaps it was because of the fiendish delight we took in missing kicking the ball, and ending up kicking the shins of our Chinese pupils. They were most anxious to learn to play; we would teach for a time, then before long they would be seen limping off the field leaving it to us. What games we had! We were ten to fourteen years old."[61]

One of the most widely held stereotypes about the resource communities of the western Canadian industrial "frontier" is that they were populated largely by "single, transient men," and as a consequence were inherently unstable and volatile.[62] Although many single men came to Ladysmith in search of employment, and possibly wives – Johnson-Cull recalls that the town's population consisted of "young married couples with large families, and young men looking for wives"[63] – single men did not dominate the town's demographic structure or community life, as they had in Nanaimo and Wellington in the 1870s and 1880s.[64] Notwithstanding the fact that by 1904 there were at least fifteen taverns and numerous boardinghouses, most of the miners lived with their families, had their own homes, and were established members of the community.[65] The household or family was the miner's main anchor in society and the community; it, as much as the workplace, class, or ethnicity, shaped working-class experience.

Men outnumbered women in Ladysmith, but the most reliable data reveal the strong presence of women and families. Statistics show that women were in the minority in Ladysmith, as in the province as a whole, and that there were more single men than married men. According to the 1911 census, the female/male ratio was 1:1.38, and 60 percent of the male population (or 1,128 out of 1,886) were listed as single while 646 men (or 34 percent of the total) were listed as married. In contrast, women made up 42 percent of the town's population, 5 percent higher than the provincial average. The percentage of single women, however, was only slightly less than that of single men. Of a total of 1,409 women, 815 (58 percent) were single, while 544 (38 percent) were married. The rest were listed as widowed, separated, or divorced. As the census data do not indicate age, and as presumably many of the individuals enumerated were children, the available figures present a somewhat skewed demographic picture.[66]

Most noteworthy in the data is the large number of married miners. Fifty-seven percent of Ladysmith's population was married and a total of 672 families were enumerated in 1911, demonstrating that families formed the backbone of the community. As Belshaw concludes in his discussion of the "frontier" demography of Nanaimo, "it was the population of married men, their wives, and their offspring who nurtured community associations, friendly societies, and even early trade unions."[67] This might not have been an adequate description of Wellington but it certainly applied to Ladysmith, where property ownership and the creation of institutions such as schools, churches, and lodges reflected the importance of family life, promoted permanence and social stability, and fostered a strong commitment to community. The alleged radicalism and militancy of BC's working class was not due to the preponderance of malcontented, itinerant single males, as has so often been argued in the past. A more likely reason was the determination of married men to secure decent livelihoods for their families. Single men, in contrast, were more likely to leave in search of other work during labour disputes or work shortages, a more difficult option for men with families, homes, and a stake in the community.[68]

Women played a crucial role in shaping the community. Life for women in small resource towns was determined by a number of important factors. Family and marital status, cultural traditions, and the level of regional economic development influenced their roles in society, as did gendered ideals of masculinity and femininity that shaped structures of authority, dominated economic production and the family, and penetrated class and ethnic relations. The ideology of separate spheres for men and women was accepted as the norm in mining and other working-class households. This attitude was reinforced by culturally entrenched preconceptions about the nature of women and their biological function and by social and economic structures. It was taken for granted that women should work in and produce for the home and family, while men worked in and produced for the market. The sexual division of labour was also maintained by the generally high wages of the miners and the conviction among miners that they ought to be the sole breadwinners. Workers in a variety of trades and industries, including coal mining, often argued that they were entitled to a "living" or "family" wage in order to provide for their families and to prevent women from having to enter the workforce, where they might compete with men for jobs and lower wage levels.[69]

Work, whether paid or unpaid but always hard and tedious, dominated the daily routine of most working-class women, although it was not recognized as contributing to the productive economy. Domestic labour consumed much of the time and energy of miners' wives and daughters. Their main task was maintaining the well-being of the family, which tended to be large, especially among lower-income immigrant families, and the well-being of the primary breadwinners – the husband, sons, or brothers. While the status of a miner's wife might traditionally have been low, her ability to maintain the household was a source of considerable pride.[70] This work, unpaid yet vital, included a variety of jobs, such as purchasing, preparing, and storing food, mending clothes, and cleaning the house, as well as looking after the children, nursing sick family members, managing the family finances, and providing moral support for the miners during times of stress or crisis.[71] Given that greasy, oily coal dust coated just about everything in a mining town and running water was often scarce, cleaning was a difficult, constant chore, made even more onerous because the Extension mines were not equipped with washhouses. This was an inconvenience to the men, who journeyed to and from Extension on the train and had to wear their wet and dirty clothes for more than an hour after quitting work, and an additional burden for their wives, who had to clean the work clothes and prepare a bath for the men.[72]

Women's influence and authority over the household economy and the raising of children could be considerable. Traditionally, women were in charge of the household finances. One English-born miner recalled that his grandmother looked after the money: "At one time my grandmother Johnstone had six sets of wages coming into the house each week. Every Saturday morning – payday – Grandmother put on a clean white apron and seated herself in a rocking chair in the kitchen where each son dropped his golden sovereigns and silver into her lap. She could neither read nor write, but she could account for every penny in the household spending."[73] The irregular nature of employment and wages in the mines, the size of the family, work stoppages, injuries, and death all had a tremendous impact on the functioning of the household. As a result, a woman's ability as a "miner's financier" took on great significance in the daily struggle for economic survival.

This role has been much celebrated in the literature, but we must be careful not to decontextualize it. The ability to make decisions about their children's upbringing, the need to supervise household finances in order to provide proper care for the family, or the participation in social groups and friendly associations must not obscure the fact that in most mining households, women occupied a subordinate position and were economically dependent.[74] They may have been "miners' financiers," but they generally had to make do with what they were given by the husband. The status and security of most working-class women, as Elizabeth Jameson has pointed out, "depended on men's incomes, health, and generosity and on the regularity of work in the mines."[75] This could be a source of domestic conflict and violence.[76] The physical and mental strain on the lives of women was such that in some mining communities, such as the anthracite region of Pennsylvania, many daughters of miners "preferred to remain spinsters rather than join their lot with mine workers."[77] This does not appear to have been the case on Vancouver Island, however, where John Belshaw has demonstrated that nuptiality was high, women married young, and "miners' daughters favoured miners' sons."[78]

Because the wages of a single breadwinner were seldom adequate for a family, women contributed to the household economy in many ways. As McIntosh and other historians have noted, working-class families were forced to develop a number of survival strategies in the struggle for subsistence.[79] One of the most common in coal towns was to take in lodgers, usually the many single men or young boys who were employed in the mines, or to operate boardinghouses. This seems to have been a more frequent practice among the women from immigrant communities, since the men often preferred to lodge within their own national group.[80] While taking in lodgers was an important source of supplementary income, it meant additional work, as women were expected to cook, clean, and mend for them. Gardening was also an important task, one that was often, but not always, a woman's chore. Many families in Ladysmith had backyard gardens or kept a few cows, pigs, or chickens, which supplemented their diet as well as their income and generally improved their quality of life. It was often the task of the woman to sell this produce, either at the town market, or door to door, as was the case with fresh milk.[81]

It was difficult for working-class women to live up to the ideal of domesticity – the notion of "true womanhood," which sought to limit women's roles to the household and the private sphere, where they would act as the moral anchor of the family and community.[82] Work outside the home was often necessary to avoid destitution during times of crisis, such as during a labour dispute, periods of unemployment or reduced hours, times of serious illness, or the death of the primary breadwinner, all of which were common features in all mining communities throughout the industrialized world.[83] Although we lack exact data for BC, an indicator of the importance of women's work can be seen in statistics from the US Women's Bureau from the early 1920s, which reveal that an estimated 20 percent of all miners' wives in the US contributed money to the household economy.[84] In the coal-mining communities of Iowa, for instance, women contributed between one-quarter and one-half of the household income; indeed, their contribution to the family income often equalled that of the husband.[85]

Women could not eliminate altogether the various dangers to the long-term survival of the household. Life for many mining families bordered on the subsistence level and was a continuous struggle. One important reason for this and for women's economic dependence was the fact that women were generally excluded from working in the primary economic sector. In Britain, women were first prevented from working in the coal mines in 1842, although their number had been declining before this date. The 1842 Mines Act, which affected only about five thousand women, was not motivated by the desire to ameliorate the arduous working conditions of women – after all, women were not prevented from working in other industries where conditions were just as squalid – but rather by the demand of urban middle-class reformers for moral improvement and greater discipline in mining communities.[86] One objective of this legislation was to force these women back into the home, where they could rediscover their "true," "natural" role – the creation of a "good" home for their husbands and children. By being "good" wives and providing a proper home, they would be in a position to combat the vice, immorality, and presumably also the radicalism that were said to plague these working-class communities.[87] Work in the mines, some reformers argued, "unsexed" women, preventing them from fulfilling their nurturing role. As one author maintains, the legislation reflected a new "ideology of physical and moral development where the practices of the pit, and any contact with them, corrupted the natural sexuality of women."[88] Another objective of the legislation, as Robert McIntosh has recently observed, was the desire by operators to reduce the authority and autonomy of the miners by weakening the hold of the family system – whereby the male miner employed his wife and children as helpers – on production. In the struggle over workplace control, women were subsequently recast as "threats to miners."[89]

Despite the efforts of the state to reconceptualize gender roles within the coal-mining population – in the words of Robert Colls, the "reformulation of women as model wives and mothers" – in the name of moral improvement, and efforts by operators to weaken the power of miners, a number of women in Britain continued to work in the mines illegally and on the surface, the alternative being destitution. According to one contemporary observer, Friedrich Engels, "the bill was adopted, but has remained a dead letter in most districts, because no mine inspectors were appointed to watch over its being carried into effect ... In single cases the employment of women may have been discontinued, but in general the old state of things remains as before."[90] Numbers declined steadily, however, and by the 1880s women constituted less than 5 percent of all surface workers,[91] the majority of whom were young girls employed at the picking tables removing dirt and slack from the coal.[92] Nevertheless, between 1852 and 1890 thirty-eight women alone were killed above ground by wagons running out of control, a figure that does not include deaths from other accidents, such as falling down shafts. For the most part, however, the ideology of separate spheres for men and women had become common sense, at least among men, by the end of the century.

This was certainly the case in the coal-mining communities of Vancouver Island. During the colonial period, First Nations women had worked in the coal yards, but this was considered unsuitable for the more cultured and civilized European women,[93] an indication of how the Victorian notion of female sexuality was not universally applied and was guided by racism. In 1877 the legislature passed the Coal Mines Regulation Act, a derivative of British regulations, and women and children under the age of twelve were prohibited from working underground.[94] While the act did not prevent women from working above ground, there is no evidence that European women did, although apparently some immigrant women from continental Europe tried to find work at the Wellington collieries.[95]

Women in the larger urban centres had more employment opportunities and even organized in the struggle for better wages and working conditions. Working-class women in small communities like Ladysmith had fewer job options outside of the home. In the neighbouring district of Nanaimo, one study estimates that in 1891 only 10 percent of the women were "gainfully employed."[96] Economic opportunities for women in Ladysmith improved as the town grew and the economy began to diversify during the first decade of the twentieth century, especially in the service industries, the public sector, and private business, but jobs for women in the local economy remained limited. As a rule, women sought work outside the home when they were still single or after the death of a husband. Such employment was seldom viewed as a long-term career but more often as a temporary occupation, and many women gave it up when they married. In some mining communities there was even a taboo on married women working.[97] The majority of women in the local labour force were young and unmarried, and those occupations where women were employed in increasing numbers – such as store clerks, teachers, nurses, midwives, office workers, and domestic help – tended to be low status and were viewed as low-wage equivalents of work in the home "that capitalized on traditional ideas of women as care-givers, or were already dominated by women."[98]

Still, there appears to have been some competition for domestic work. During the commission into Chinese and Japanese immigration in 1902, one miner argued that the Chinese competed unfairly with white women. "Domestic service," he stated, was "a large sphere where women may earn an honest livelihood and learn to fit themselves for more important duties in life."[99] While such jobs may have enabled women to earn a livelihood and to learn future domestic skills, they were rare in Ladysmith and did not allow women to become economically and socially independent. If there was an exception to this rule, it was the small number of independent middle-class businesswomen who owned and operated a variety of stores and services, including a hotel, general stores, and several specialty shops.

Women's subordinate position in society was also reinforced by attitudes rooted in the male-dominated workplace that fostered rigid stereotypes of masculinity and femininity, inevitably to the disadvantage of women. It is often argued that the specialized language of the mines, the masculine "pit talk," created barriers between the men initiated into the coal-mining culture and the women who lived forever "on the surface."[100] The job itself "socialized" men "into manhood," teaching boys such traditional "masculine" values as independence, courage, toughness, co-operation, and common interest, views perpetuated by trade unions.[101] Women, in contrast, embodied a different set of virtues. They were nurturing and caring, yet irrational and weak. Socialized to be wives and mothers, they were consequently excluded and protected from this masculine, work-based proletarian culture. Attitudes towards gender influenced relations in other ways. For a miner to lose his status as breadwinner through lack of work or injury was a serious psychological, not to mention financial, blow to the man. Incapacity through illness or injury was often considered emasculating, and reliance on the income of a wife or daughter during times of crisis, such as a strike, could be a source of considerable depression and have an adverse effect on strikers' morale and solidarity.[102]

The effects of such patterns of gender identity and socialization may be verifiable, but the extent to which women's exclusion from the male-dominated culture of the mines isolated and marginalized them is unclear.

The fact that in coal-mining communities in general miners' daughters tended to marry within their class suggests, among other things, a high degree of identification with the cultural baggage of their partners.[103] Certainly women's understanding of the working conditions and culture of the mines, particularly the impact of the work environment on male family members, was considerable, even if it was not as profound as the men's intimate knowledge gained through the immediate experience of work, and even if "miners sheltered their families and especially their wives from the worst that went on in the mines."[104] Whether in Cape Breton, Scotland, northern France, or Ladysmith, women, as much as men, felt the consequences of the strikes, the periods of unemployment, and the numerous injuries, illnesses, and deaths, all of which added to their already substantial labours and stretched their economic and emotional resources. Indeed, because the mining household was dependent upon the earnings of the breadwinner, wives, as "miners' financiers" or "household managers" had, in the words of one recent author, "a particularly immediate connection to workplace issues."[105]

Women in the resource communities also played a crucial role in the development of community life, the creation of a sense of common identity and interest, and the functioning of the local economy. They were active in the public domain, especially in the various religious and social organizations that proliferated in Ladysmith. Vancouver Island coal-mining towns, as in Britain, Europe, and throughout North America, had a rich tradition of associational life. The various church groups, fraternal organizations, mutual aid societies, ethnic lodges, and social clubs were vital to the community, providing health and accident insurance and fund-raising for numerous charities, and offering a wide range of social events and activities.[106]

Most of the societies, clubs, and lodges were organized according to a sexual division of labour and leisure. Women's branches or auxiliaries of the principal associations were set up in Ladysmith for social and philanthropic purposes, such as the Laurel Rebekah Lodge No. 9 of the Independent Order of Oddfellows, which had been established in Wellington in 1895 and moved to Ladysmith in 1901, and the Pythian Sisters Temple No. 5, the women's branch of the Knights of Pythias. Such groups were a valuable vehicle for participation in the public sphere and shaped community life. By their very nature, however, women's auxiliaries reinforced gender divisions and reflected women's subordinate status in society. As

Joan Sangster and Linda Kealey have pointed out, women were expected to perform "female" tasks within these organizations, tasks that did not challenge male prerogative and authority, and they rarely had any influence over the formulation of policy.[107]

Although the active associational life in Ladysmith reinforced rather than undermined established gender relations, many women became prominent members of the community through their leadership roles in the auxiliaries. Through their work in the many different service clubs, women functioned relatively autonomously and independently in the public sphere and exerted considerable influence over decisions affecting the day-to-day life of the community without always having to compete with or be subordinated to the demands of men. Membership in these associations also crossed class lines. While the United Mine Workers Women's Auxiliary was socially exclusive, membership in other groups and associations in Ladysmith tended to be inclusive. Many of the friendly societies, clubs, and philanthropic organizations drew members and leaders from the ranks of businesspeople and the working class alike. The Ladysmith Hospital Auxiliary, to cite just one prominent example, was organized and run by the wives of the miners and merchants. One of the most important societies in the town, the Hospital Auxiliary provided essential services to the new hospital, built by the miners and administered by the Miners' Accident and Burial Fund. It was also very active in assisting injured or sick workers and the widows and orphans of miners killed in cave-ins and in the explosion of 1909, and held numerous charity events to raise money for supplies and equipment.[108]

That membership in the Hospital Auxiliary and other friendly societies crossed class lines is of considerable significance. Local chronicler and historian Viola Johnson-Cull recalled that "there were two distinct types of society in Ladysmith, those that worked in the mines ... and all the business people ... The business people didn't fraternize with the miners."[109] This is clearly an exaggeration. Although social relations were determined by a mix of class, gender, and ethnicity, and while racism was as prevalent here as in other BC communities, Ladysmith was not a polarized, conflict-riddled, unstable society on the industrial frontier. While social divisions were very real and confrontation a constant possibility, organizations such as the Ancient Order of Foresters or the Ladysmith Hospital Auxiliary helped to build a strong sense of community that mediated division. In other words, they served as vehicles in the process of negotiating social relations and defining one's place in the community. Indeed, this brief survey reinforces the fact that social relations are inherently complex and that community is a historical process. There was a high degree of flexibility and accommodation, no doubt a reflection of the predominance of coal miners and their families in the town, and of the ethnic, religious, and social diversity of the town's inhabitants.

# 3

# *Down in the Dark and Gloomy Dungeons*

The growing diversity of Ladysmith's economy and the emergence of a more cohesive community must not obscure the fact that the Extension colliery dominated the economic life of the region. Similarly, work in the mines – or lack of it – dominated the daily life of the miners. Family and community shaped the contours of working-class experience in Ladysmith, but it was the job itself that ultimately defined the identity of the miner. Here, at the sharp point of production, where the harsh reality of work underground was confronted by the exigencies of capitalism, was the true source of class conflict and socioeconomic inequality. It was also here, at the sharp point of production, where a potential means of emancipation was to be found, for if anything was to forge class solidarity and consciousness – to provide the disposition to act as a class – then it was the process and conditions of work.

In 1911, the year before the Great Strike of 1912-14, the Extension mines employed a total of 881 men. The vast majority of the men, about 714, worked underground, producing 331,576 tons of coal during that year.[1] The colliery was a large, modern industrial enterprise, the heart of which consisted of four separate mines that tapped into the vast Wellington seam. Three of the mines (No. 1, No. 2, and No. 3) were connected and worked from the main rock entry tunnel, which was driven at a slight upward angle into a hill through the rock underlying the coal seam. From the pithead, the miners entered the main tunnel, which extended approximately 5,200 feet, until they reached a landing where the tunnel hit the seam and the three slopes that led to the different working levels of the mines. No. 4 was a separate mine located along the railway line about one and a half miles from the main tunnel and was opened by a shaft 290 feet deep.

Extension miners, c. 1904.
*BC Archives, E-02581*

Several methods were employed to extract coal at Extension, and indeed in mines throughout the world at this time: pillar and stall (also known as room and pillar or board and pillar) and longwall mining. Another technique, the extraction or robbing of pillars, was, as its name suggests, a part of the pillar and stall process. In pillar and stall mines, parallel sets of cross-entry tunnels or levels were driven at right angles from the slope or the main entry shafts. Crosscuts were then dug every twenty or so yards to connect the cross entries. From these cross entries a grid of stalls was mined out, separated by solid rectangular pillars of coal about sixty feet wide that were left in place in order to support the rock roof or overburden. The deeper the mine, the more difficult pillar and stall became because of the increased pressure exerted on the overburden; as a result, larger pillars had to be left in order to compensate for the weaker roof. The stalls, the workplaces at the face of the seam, varied considerably in size, depending on the thickness of the seam. Although some stalls in the No. 2 mine at Extension, for instance, were twenty-one feet wide and as high as twelve feet, more often than not the seam was shorter than the miner, making the coal difficult for the miner to dig in the uncomfortable, confined space.[2]

Longwall mining was a completely different process.[3] It exposed the entire face of the seam and was sometimes worked by larger numbers of men, either outward from the main slope (longwall advancing), or backward from tunnels driven to the boundary of the seam towards the slope (longwall retreating). In this way, miners removed a greater portion of the seam, making the process more efficient and cost-effective. In place of pillars to support the roof, the miners erected a system of temporary supports parallel to the working face. These timbers would take the weight of the settling roof but leave enough space to undercut and remove the coal. As the face receded, the supports would be moved forward. The roof of the mined-out area (the gob or goaf) might then settle on its own, or the gob would be backfilled with waste.

The choice of mining technique depended on a number of factors, including geological structure, the width and the thickness of the seam, the depth of the mine, and the capital cost of development. In some cases, operators tried to adjust the method of extracting coal as the conditions underground changed. Pillar and stall, the traditional method of mining coal, was favoured with wide seams and was ideally suited to shallower pits; longwall mining, which was used in the later and deeper mines, was more suited to hard, thinner, and more uniform and fault-free seams. Because the seams on Vancouver Island were highly fragmented and varied in thickness, pillar and stall predominated, although both techniques were sometimes used in the same mine, depending upon geological conditions. In the Wellington No. 5 pit, both systems were often employed, as was the case at the Vancouver Coal Mining and Land Company's (later the Western Fuel Company's) Protection Island shaft, where in 1903, for instance, the upper seam was mined by pillar and stall, the lower seam by longwall. Likewise in 1911 both methods were used in the No. 2 mine at Extension.[4]

Longwall mining offered operators a number of potential advantages. It allowed narrow seams to be mined profitably because a higher percentage of the seam was removed, yielded more of the valuable large or lump coal, which was more profitable, and provided more continuous production. Longwall was also thought by some to be a safer method of working the coal than pillar and stall, although it presented a different set of safety problems. Gas, a continual hazard in all operations, and roof falls posed unique challenges in longwall because so much of the face was exposed. Along the slope of the Wellington No. 5 mine and some districts of No. 6, where longwall was sometimes used, high levels of gas were occasionally produced by roof falls. The risk of methane could be reduced with proper ventilation; it was much easier to direct air along the face of a longwall than through the maze of tunnels in a pillar and stall district.[5] Finally, although longwall mining could create additional expenses – largely because of the cost of backfilling and maintenance – the technique potentially lowered working costs because it reduced the amount of timbering, kept haulage roads to a minimum, and required fewer on-cost day labourers, such as tracklayers and drivers. In addition, the enhanced revenue generated by the increased production of profitable lump coal added to its benefits.

Some mining historians have attributed great importance to the choice of technique. Longwall, they argue, was not just a practical decision about how the coal was to be mined; more importantly, it revolutionized production, structured the workforce in fundamentally different ways, and determined social relations. Instead of small, isolated, and relatively autonomous teams working in stalls, the longwall technique required larger numbers of more easily supervised men working together at the face. According to Priscilla Long, it brought "the underground workplace a step closer to the factory system."[6] John Douglas Belshaw has made a similar case for the mines of Vancouver Island, arguing that longwall mining introduced competition among workers, rather than cooperation; allowed the Dunsmuirs to replace skilled European colliers with unskilled Chinese and Japanese labourers, creating a racial and social division of labour; enabled management to supervise more actively the workforce at the face; and facilitated the introduction of machines underground.[7]

By itself, however, longwall mining did not revolutionize production, transform mines into factories, erode the skills of the practical miner, or accelerate the process of proletarianization.[8] Prior to the introduction of machine mining, the organization of work on the longwall did not represent a radical departure from traditional pillar and stall mining. Longwalls were generally mined by small crews of four or five men, and while the potential for greater supervision existed, the autonomy of the crew system was preserved. As Roy Church concludes, in Britain "longwall teams still worked under their elected leaders, and men habitually refused to accept supervision and advice from deputies, refusing to be 'spied upon.'"[9] In fact, longwall may have strengthened rather than weakened the autonomy of the crew system, and hence the position of practical miners. As

Keith Dix and John H.M. Laslett have pointed out in their studies of the history of US coal mines, longwall, especially after the introduction of large-scale team mining, often empowered the miners as it gave them a strategic advantage in their disputes with management that did not exist with pillar and stall mining. "By making the extraction of coal dependent on the simultaneous presence and co-ordinated activity of thirty or forty men," writes Laslett, "the absence of two or three of them could bring production to a halt in a manner not possible under the 'room and pillar' method."[10] Because successful longwall mining depended upon highly coordinated, uninterrupted production, interruptions could cause the roof to become unstable or even to collapse, necessitating additional cost to re-establish the working face. Stoppages could therefore cause serious, expensive damage to the mine.[11]

At the same time, miners themselves did not feel that the introduction of longwall eroded their skills or status. The same skills used in pillar and stall were required at the face of a longwall, and the technique reinforced the need for cooperation and teamwork rather than creating a new competitive edge, because it required a more experienced, disciplined workforce in order to integrate and coordinate the underground systems of production.[12] Miner resistance to the introduction of longwall, while not uncommon in Britain, where the technique spread rapidly between 1860 and 1880, came not from the fear of proletarianization but rather from the fear that they would not be adequately remunerated for the more intensive work. Church concludes that "enhanced earnings one way or another usually proved to be the solvent of hostility."[13] This was clearly the case in February 1903, when miners at Nanaimo successfully struck for an increase in the rates for lower-seam longwall work.[14]

Belshaw claims that there was no "steady evolution towards one system at the expense of the other,"[15] but his argument is based on the false assumption that there was indeed an evolution towards longwall mining and that it prevailed in the Dunsmuir mines. However, longwall mining, despite its apparent advantages, was used only sparingly on Vancouver Island before the First World War. There is no evidence to suggest that the decision to use longwall on Vancouver Island was made for reasons other than the thickness of the seam. Pillar and stall (followed by robbing the pillars) was the dominant method of extracting coal in the Island mines, including those owned by the Dunsmuirs, and in North America generally during the nineteenth and early twentieth centuries, and in fact continues to be used today.[16] This traditional technique was much more suitable to the geological and production conditions on Vancouver Island, because it was less capital-intensive than longwall and was a much more flexible method of mining the geologically disturbed regions of the Island, since longwall was feasible only if the seams were narrow, relatively free of faults, and not very steep.[17] Rather than being used almost exclusively in Dunsmuir's mines, longwall was used in the VCMLC's operations on Protection Island and at Northfield, where the seams were very narrow, and at the small East Wellington colliery, where longwall was the only profitable way of winning coal and competing with the Dunsmuirs and the VCMLC.[18] Similarly, in Dunsmuir's Union No. 7, where in the early twentieth century high-grade anthracite coal was mined by machine, the seam was narrow – only three and a half feet – and longwall was the only way it could be worked. In fact, in 1897, to cite just one year, of the five pits at Dunsmuir's Wellington operation, only one was longwall (No. 1); No. 5 used both longwall and pillar and stall, and No. 3, No. 4, and No. 6 were pillar and stall.[19] Most of the work at Cumberland and Extension during the 1890s and the early twentieth century was either pillar and stall or the extraction of pillars. Of the seven mines at Cumberland, only No. 7 used the longwall method; No. 5 used both techniques and the others employed pillar and stall.[20]

Because pillar and stall remained the dominant form of mining coal, the technique of pillar extraction or robbing the pillars – in northern England this was also referred to as "judding" or "drawing the jud" – was widely employed in the mines of Vancouver Island, although it is not mentioned at all in Belshaw's studies. Robbing the pillars, which had been developed in England in the 1830s by the mine operator and inventor John Buddle, was designed to increase the yield of coal in pillar and stall operations.[21] One of the main benefits of longwall mining was that it could theoretically remove 65 to 80 percent of the coal in a seam.[22] In pillar and stall mines, often less than 50 percent was removed, leaving large amounts of coal unmined in the form of pillars. A 1914 report estimated that as much as two-thirds of the coal at Extension was left as pillars.[23] Mining the pillars once the stalls had been worked out brought the extraction yield much closer to that of longwall. Removing old pillars at Extension, for example, could result in an extraction of 50 percent of the existing coal, removing new pillars, 90 percent.[24] Consequently, with the exception of exceedingly narrow seams, there was little incentive to use longwall

Miners entering Extension mine in coal cars, c. 1905.
*BC Archives, E-01184*

because the companies were able to compensate for the low extraction rate in pillar and stall mines by robbing the pillars.

The extraction of pillar coal normally began shortly after the rooms were abandoned, but operators sometimes had to wait a considerable period of time, because roof subsidence could disrupt the entire operation underground. Robbing the pillars was one of the more dangerous jobs in mining. In the worked-out districts of the mine, known as the "broken," the air was often worse, gas accumulation was high due to compression from the weakened overburden, and the risk of roof falls was greater. According to a 1933 study by the US Bureau of Mines, fully one-third of all roof-fall fatalities occurred during the extraction of pillars.[25] Because of the risks and skill involved, however, pillar men were sometimes paid more and the work was coveted by experienced colliers.[26] At Extension, as the miners cut down the pillars, they often supported the roof with "shanties" or "cogs," square wooden-framed structures built "log cabin style" that were filled with rocks and waste and were sometimes as large as four feet by eight feet wide, depending on the condition of the roof. Eventually, the mined-out district was abandoned altogether and left to settle. While some contemporaries condemned robbing pillars as a "reprehensible practice" because of the dangers involved, in most mines it was the only way to mine coal efficiently and effectively; the operator who did not rob pillars was thought to be a wasteful and careless businessman.[27]

Mining the coal, transporting it to the surface, and preparing it for distribution constituted a complex operation requiring a large, coordinated labour force. Ian McKay describes the mine as "a dynamic machine for winning coal, a moving complex of systems – those of production, transportation, distribution, maintenance and management – each one of which was required for the functioning of the whole."[28] This system of production was governed by a highly structured and hierarchical division of labour. The colliers or hewers of coal, also commonly referred to simply as miners or practical miners, constituted the bulk of the underground workforce. Of the 714 men who worked underground at Extension in 1911, 360 were classified as practical miners.[29] They were responsible for production; in effect, they were the heart of the mining operation. The function of the other employees was to ensure that the miner could produce coal.

A miner poses for the camera in his stall at Extension No. 2.
*BC Archives, E-02767*

The job of winning coal from the face and loading it into cars was done by hand. At Extension and elsewhere, the production of coal centred on small teams or crews, usually consisting of one or two experienced colliers and their helpers, who worked by the light of an open-flame lamp at the face of their assigned stall. Miners often developed a strong attachment to their stall. They knew how its roof worked, the contours and peculiarities of the seam, and devised the best means of winning the coal. Sometimes stalls were even named after the miner, and during absences from work they were often kept for the miner upon his return. At Extension, for instance, John Wargo's workplace was known as "Wargo's Slant." Likewise, a long heading dug by Tully Boyce's crew at Nanaimo was called "Boyce's Incline." According to one contemporary observer, "a miner is as attached to his stall, as many a lady is to her drawing room."[30]

Once the miner had made the long, arduous journey from the pithead to the working face, his work began. The miner had to perform a number of different tasks, often in physically confined space that was as "black as Walpurgis night,"[31] and in a dusty, hot, and repressive atmosphere devoid of sufficient oxygen. Each task required considerable skill and knowledge. His first job upon reaching his place was to use his pick to make two vertical cuts or shearings at each end of the face (sometimes shearings were not made; this was known as a "shot off the solid"). He then made a three- to four-foot deep undercut, often about a foot in height, at the base of the seam. The miner did this lying on his side and, as work progressed, actually underneath the overhanging coal, which he supported by short blocks. Undercutting was time-consuming – often taking up a better part of a shift – and physically demanding, requiring considerable dexterity with the pick.[32] After completing the undercut, the miner had to remove the coal from the face. Before the use of explosives became common in the second half of the nineteenth century, this was done by hammering wedges into the top of the seam, forcing the coal to drop.[33] Powder was introduced into the Island mines in the 1860s; at Extension the coal was blasted from the face. Using an auger, the hewer and his helper (one man held the bore, the other hammered it) drilled a long hole the depth of the undercut and then inserted a plug or cartridge of blasting powder, tamped it with clay, broken shale, and even highly dangerous coal dust and fine coal, and put in a fuse. After he lit the fuse, the miner left the stall as quickly as possible; if the shot was fired correctly, the subsequent explosion would bring down the coal.

Blasting was a highly skilled job. The miner had to judge carefully the amount of powder required. If he did not use enough, the coal would not break from the face. If he used too much, the coal might shatter into small pieces or, worse, the blast might dislodge the timbers, causing a cave-in or an explosion. Blasting was also highly dangerous; there was great concern about using powder in mines where safety lamps were required, because the fuse was lit by an open flame. In BC, as throughout North America and Europe, procedures were introduced to minimize the danger of blasting. The Coal Mines Regulation Act of 1877 (CMRA) required that firing be done "by or under the direction of a competent person who shall be appointed for the purpose."[34] The regulation, however, was sufficiently vague to allow considerable room for manoeuvre. Thus, in some mines only shotlighters were permitted to fire shots, which they usually did between shifts after determining the number of shots each miner needed to load coal for the day. In other mines, shotlighters examined the drilled holes, the quantity of powder, and the way the holes were tamped before permitting the men to fire. Often, however, the miners did not wait for the shotlighter; they resented the devaluation of their skill and knowledge that the rule implied and often literally could not afford to wait for a shotlighter.[35]

While the collier in his stall constituted an autonomous unit of production, his work was made possible by a large body of workers who performed an extensive range of jobs. These on-cost employees were essential to mining operations and consisted of men and boys who worked above and below ground. According to a commission of inquiry into the provincial coal-mining industry, the on-cost underground workforce in an average BC mine was normally composed of "firemen, shotlighters, bratticemen, timbermen, timbermen helpers, tracklayers, roadmen, driver-boss, drivers single, drivers double, pushers, winch-drivers, rope-riders, doorboys, stablemen, pumpmen, linemen, machine-runners, drillers, muckers, cogmen, pipemen, loaders, motormen, cagemen, and other labourers." The workforce above ground was almost as extensive: "blacksmiths, blacksmith helpers, machinists, machinists' helpers, carpenters, carpenters' helpers, engineers on the slope or shaft, engineers on compressors, firemen, teamsters, tipplemen, and other labourers." Not surprisingly, it often cost the companies more in wages for these on-cost workers than for practical miners.[36]

Extension miners, boys, and mules, c. 1905.
*BC Archives, E-01185*

Many of the underground positions were held by men and boys learning the trade. Miners often hired their young sons or relatives as helpers. Here, beginning around the age of twelve, they served an informal apprenticeship before rising up the job hierarchy to practical miner or beyond.[37] Sometimes they started as doorboys or trapperboys, opening and closing ventilation doors underground, before advancing to mule driver. Older boys might work alongside their fathers as helpers in the stalls, doing a variety of tasks such as loading the coal into cars, timbering the stall, or removing waste. Helpers or apprentices were paid by the miner, not the company, out of the total daily earnings of the team.

The employment of boys in mines was traditionally an integral part of the process and economy of mining. Not only did the boys learn all aspects of the job and provide companies with a trained workforce for the future but their labour was an important "survival strategy" for mining families in the struggle for subsistence.[38] On Vancouver Island, the number of boys working in the mines was small. In contrast to Nova Scotia, where in 1900 boys constituted over 12 percent of the workforce, on Vancouver Island in only one year, 1896, did they make up more than 5 percent.[39] A number of factors mitigated against the employment of boys in the Island mines. The recruitment of young boys was made difficult because the CMRA restricted the age at which children could enter the mines. No boys under twelve years were permitted to work above or below ground, while boys under sixteen were allowed to work a maximum of only thirty hours per week underground. The availability of unskilled Asian labour also limited opportunities for boys and also enabled management to undermine the miners' "family-based organization of labour," which "committed the miner 'to the ideal of the hereditary closed shop.'"[40]

The CMRA's restrictions on child labour and the use of Chinese workers were controversial, as they made it harder for families to make ends meet and for boys to find work. And indeed, wage demands figured prominently in most strike actions.[41] The miners also resented interference in their traditional right to hire helpers of their own choice. Although Francis Little, general manager of the Wellington Colliery Company, claimed in 1902 that "the white man can take on as helper whom he pleases,"[42] many, including Samuel Robins, believed that the Mining Act and especially the employment of Chinese and Japanese workers reduced opportunities for miners' sons. In Wellington and Cumberland, where Dunsmuir permitted the Chinese underground for lower fixed wages, white miners often hired Chinese labourers to perform jobs that might have been done by their sons. The resentment of some white miners can be seen in testimony provided by the miner Aaron Barnes. When discussing James Dunsmuir's proposal to settle the 1903 strike – which in effect amounted to a 20 percent wage reduction – Barnes commented: "But in the event of the acceptance of a proposition of that kind, the company furnishes the helper to each individual, and it is not said as to what kind of a helper that might be in this proposition. He might be one of those long-tailed gentlemen."[43]

The apprentice or helper was usually responsible for bringing the empty cars up to the coal face and also helped to shovel the coal. Once they had loaded the coal, tagged the cars with the miner's identification number, and pushed them out to the main entry, the cars were brought to the collection point at the bottom of the slope by mule and then connected together to form a long trip for transportation to the surface by the electric motor hoist. The job of mule driver was a difficult, strenuous occupation,

Electric train pulling trip of coal to the surface at Extension.
*BC Archives, E-02770*

often reserved for boys sixteen years of age or older. Despite enormous technological advances in underground transportation and haulage, mules and ponies remained indispensable to mining operations well into the twentieth century. The driver was critical to underground production because it was essential to remove the mined coal from the stall as quickly and efficiently as possible, and, in order to make maximum use of production time, to bring empty cars back to the levels when the miner needed them. Lack of boxes meant lack of wages, and disputes about shortages of cars or delays were common.

At the tipple or pithead, the tags on the coal cars were recorded and the coal inspected for impurities and weighed. These latter tasks were of considerable importance, but were also a source of discontent to miner and company alike. Although the miner was supposed to load only clean coal, some places were dirtier and rockier than others, and it was difficult to avoid including slack in the car. The company, however, had to bear the cost of the waste, "not only in the initial price to the miner, but also in their proportion of the cost of all the other handling in the mine and at the tipple, and the full cost of picking, washing, and of making away with it."[44] Consequently, management tried to regulate this. Too much waste in the car could result in "docking," the confiscation of the car, and even the dismissal of the miner.

Pit boss and crew posing at the cross-over dump at Extension, c. 1908.
*BC Archives, E-02768*

Sorting coal at the picking table, c. 1910.
*BC Archives, I-31976*

After inspection and weighing, the coal was unloaded onto a conveyor belt and delivered to two sets of parallel shaking screens that separated the lump coal from the small coal. At Extension the oversized lump coal was conveyed to the picking table, where Chinese labourers and young boys removed further impurities. Only then was it dumped into railway cars and sent to the docks at Ladysmith, where it was either loaded on ships or sent by barge to the mainland. The small coal, approximately 40 percent of total output, was also shipped by rail to Ladysmith, where it was dumped into the washers to be sluiced, sorted into two grades (nut coal and pea coal), and stored in huge bunkers that dominated the shoreline. About 34 percent of the screenings that passed through the washers was waste.[45]

Mining coal was like few other jobs in the industrialized world. The miner laboured hundreds of feet underground in conditions that were always dark, dirty, damp, hot, and cramped. Devoid of natural light, fresh air, and physical space, it was an unnatural and unhealthy environment, one totally alien to human beings. The miners' troglodytic existence was compounded by an extraordinary host of perils that endangered their lives. Explosions, rock and coal falls, and other accidents killed, maimed, and injured more workers in the coal mines than in almost any other industry. As if this were not bad enough, a miner's life was often cut short by disease, the most common of which was miner's lung, or pneumoconiosis; many also suffered physical deformities from working stooped for hours, or from their injuries.

The organization of work was also unique. It is noteworthy that regardless of the geological structure of the coalfield or of developments in technique, in level of mechanization, in social relations, or in the ethnic composition of the workforce, the work process was characterized by remarkable continuity rather than by radical change. Although coal mining fuelled the industrial age, "the coal mines had escaped – as late as the 1920s – some of the characteristic effects of industrialization." Mines required large amounts of capital investment and were heavily mechanized and technologically advanced industries, but "from the depths of the coal mines, where the great majority of coal miners worked, the industry resembled not so much a factory as the scattered settlements of a rough, primitive, and often pre-industrial countryside."[46] Rather than being modern factory operatives, practical miners were like subterranean artisans; their stalls were their workshops underground. They were the most skilled and experienced workers in the mine. Like master craftsmen, they possessed almost complete knowledge and control over the pace of production and regarded themselves as autonomous producers. They, rather than the foreman, exercised the most control over the work process and worked with considerable autonomy at the face; they supervised and often subcontracted on-cost labour and helpers; and they determined their own level of production, comfort, and safety. In addition, like preindustrial artisans, they owned their own tools, were paid not by the hour but by the ton (that is, according to what they produced), and sought to regulate access to the trade through an informal apprenticeship structure and through control over hiring practices.

The practical miner's skill defined his status and was the cornerstone of his strong sense of independence, tradition, and collective identity. However, these skills were seldom acknowledged by the general public because the finished article – a lump of coal – was not "recognizable as a product of skilled labour."[47] As Royden Harrison notes: "The bourgeois gentleman might as well warm his posterior before a fire made up of small coals as of large."[48] In fact, the skills of the miner were highly contextualized and encompassed both subjective and objective knowledge. First, the miner acquired a series of work-specific skills, skills he required to dig coal. These included not only the ability to swing a pick, to erect timbers, and to wedge or to blast with powder but also the ability to read the coal face: to determine how best to undercut or shear, how best to fire a shot in order to produce a given quantity and quality of coal, and how best to load it in order to reduce breakage and to save his energy. Second, the skilled miner learned survival techniques. Perhaps more than in any other job, the miner depended upon his wits and senses. He had to be able, in McKay's words, "to watch out for the dangers which always surrounded him, to listen for the least noise, to smell an abnormal odour, to spot a poorly timbered roof, to notice the unusual appearance of a lamp or the strange sound of a pick."[49] In many respects, these intangible and little-appreciated skills were his most important, as his life and the lives of his coworkers depended on them.

Because production was dependent upon the skills of the collier, operators had great difficulty rationalizing the workplace and introducing greater supervision and discipline. Unlike most manufacturing industries at the turn of the century, the coal mines remained relatively impervious to the demands of scientific management. Similarly, mechanization, a key element in the deskilling and proletarianization of factory labour, did not necessarily reduce the need for skilled workers in the mines.[50] Although coal-mining operations witnessed tremendous technological advances during the course of the nineteenth century, these developments had only a limited impact on the underground work process. The mines of Vancouver Island were never as extensive as those in Europe or the United States, but they were technologically advanced. It has been suggested that mechanization progressed most rapidly during the Great Strike of 1912-14.[51] In fact, the development of the coal mines on Vancouver Island was accompanied by the use of technology, the first steam engine arriving at Fort Rupert in 1851.[52] As the industry expanded in the 1870s and 1880s, steam-powered water pumps and ventilation systems were quickly introduced in Wellington and Nanaimo, enabling mines to be dug more deeply and production to increase. Improvements in haulage and hoisting followed. In early 1891, an Edison General Electric generator, driven by a steam engine, was first installed at the Nanaimo colliery to power the four electric locomotives that were used to transport coal to the surface. In addition, the two miles of cable underground supplied "the bottom of the shaft, sidings, and engine houses with light in the form of the electric spark."[53]

John Belshaw has argued that the general availability of cheap labour meant that "the incentive to mechanise the Dunsmuir collieries was low."[54] On the contrary, all major operations on the Island were very keen on mechanization. According to one contemporary observer, both John Bryden and Francis Little, senior managers for Dunsmuir, held "advanced views as to the adoption of every possible mechanical means for economizing labour and time."[55] The problem was that new technological developments did not necessarily improve actual production. The mechanization of the pumping, ventilation, and haulage systems made these operations safer and more efficient in terms of both labour and cost, but they did not have a dramatic impact on the extraction of coal or the organization of labour underground, since the digging of coal was still done by hand. This changed with the introduction of the undercutting machine, which came into sporadic use in the United States in the 1870s. The first undercutting machine on Vancouver Island, a "chain-breast" undercutter manufactured by the Jeffrey Manufacturing Company of

Columbus, Ohio, began operating in the No. 4 slope of the Comox colliery in 1891. By 1894 that number had increased to four. A decade later, compressed-air machinery was being used to mine and drill in the lower seam of the VCMLC's Protection Island shaft. Although the undercutting machines were best suited to longwall mining where there was a continuous seam, the machines used in Union No. 4 worked in stalls.[56]

In 1891 the provincial mine inspector described with great enthusiasm how the undercutter functioned:

> They do good work, and, what is about the hardest to do, viz., undermining, this being the work that they are made for. The machines, when at work, stand end on to the face of the coal, with the cutter bar three feet three inches long; this at regular distances from each other is set with teeth, so that there is four inches cut, neither more nor less. After the machine is set in motion, in four minutes it has undermined a hole close to the floor three feet three inches wide, six feet in the face and four inches high; when it has done this work it comes out again, and then it is put over for another cut, which takes about five minutes, and so on until the place is all undermined, when a shot in each corner will bring all the coal down that is undermined. About one of the best day's work that one of these machines has done here was to undermine ninety feet long and six feet deep; this must have been a great saving in powder and coal, not to say anything in the saving of labour.[57]

The undercutting machine clearly offered numerous advantages to the operators: they eliminated the labour-intensive and time-consuming task of undercutting and promised to increase output. More significantly, the new machines had the potential to reorganize the workforce to the advantage of management, by reducing its dependence upon skilled labour. This was recognized by the Illinois Bureau of Labor, which in 1888 reported that not only did the machine lead to a division of labour but it relieved the operator "for the most part of skilled labor and of all the restraints that implies. It opens to him the whole labor market from which to recruit his force."[58] As a result, many people predicted that there would be great labour unrest if operators used machines as a weapon in their struggle against organized labour.[59]

Despite these potential benefits to the operator, Robins and James Dunsmuir did not attempt to replace their colliers with machines and did not see them as a means of breaking the power of the skilled labour.

In contrast to the United States, where 24.9 percent of coal was machine-mined in 1900, in 1915 only 10.43 percent of the provincial coal output was mined by machine, and 89 percent of that total was produced by the Western Fuel Company in Nanaimo.[60] Only in 1930, after the onset of the Great Depression, did coal-face mechanization increase substantially.[61] Consequently, we must be careful not to exaggerate the extent to which mechanization deskilled and proletarianized the mining workforce. In some situations where machines were introduced, it was less a question of deskilling than reskilling. As David Frank has pointed out in his study of Cape Breton mining, even though machinery was accepted over time at the face, it "only partially affected the organization of work ... At the coal face and on each level, the number of men employed in hand-powered operations greatly exceeded the number of those running machines."[62]

Why was James Dunsmuir reluctant to take advantage of the undercutting machines to reduce the influence and authority of skilled labour? A number of historians have argued that the availability of cheap Chinese labour made mechanization unnecessary.[63] But the Chinese worked underground only at Union, and were not employed underground at Wellington after 1889 or at Extension. Furthermore, the first machine was used at Union. A number of factors may have inhibited their spread on Vancouver Island. In the first place, the capital cost of the new equipment and the infrastructure needed to support it was substantial and could entail a high level of financial risk for operators. Second, successful use of undercutting machines depended upon geological conditions. A soft roof or steep inclinations of the seam made the machines less effective. As one contemporary reported in 1900: "The mines at the [Pacific] coast are not so favourably circumstanced as those in the east for the extensive use of coal-cutting machinery owing to the irregularity of the pavement and the 'faulty' character of the seams, and the tonnage produced in this manner is not large."[64] Third, the new machinery posed a threat to the health and safety of the miners. They vibrated terribly, were noisy, often causing their operators to become deaf, and prevented the miners from listening to the working of the roof, which made them impractical for the dangerous task of robbing the pillars. They also increased the amount of coal dust and consequently the possibility of explosions. Fourth, machine mining was destructive of the highly valued lump coal produced in the Island mines.[65] Finally, while the mining machine could undercut the coal faster than a pick miner could, there is evidence to suggest that machine mining was not always more productive than mining by hand. The reason for this apparent anomaly was that regardless of how quickly and efficiently the machine undercut the coal, the coal still had to be manually loaded into the cars, a labour-intensive, time-consuming process. As Dix has pointed out, production would be truly revolutionized only when mechanical loading and improved haulage systems were introduced in the US in the 1920s, and on Vancouver Island in the 1930s. Only then, he concludes, could the coal-mining industry "realize substantial gains in efficiency and profits."[66]

Mechanization of mine operations on Vancouver Island was rarely if ever a source of conflict and tension.[67] Most of the improvements in ventilation, haulage, and lighting directly benefited practical miners as well as the on-cost workforce. In addition, few men complained about the introduction of cutting machines. The main source of discontent came from the operators' attempts to reduce labour costs, their primary method of ensuring production and profits. On Vancouver Island, this was achieved through general wage reductions, seasonal layoffs, lockouts, and the employment of cheaper Asian, particularly Chinese, labour.[68]

The employment of Chinese miners on Vancouver Island began in 1867 when a small group was hired by the VCMLC. The employment of Chinese labourers at half the wages of the European workers was initially regarded as beneficial to the industry. Operators were able to reduce the cost of labour at a time when labour was scarce; white miners were able to profit by subcontracting work to the Chinese as mine helpers.[69] As long as the Chinese did not directly threaten European miners' wages or jobs, they were generally tolerated. The first major confrontation came in 1870, when the VCMLC attempted to use Chinese workers as strikebreakers during a wage dispute. There was an outcry not only from the miners but also from the general public. According to Gallacher, "even Victorians voiced strong opposition to the scheme, claiming outright that the fabric of colonial society would be at stake if Orientals were allowed to replace whites in the labour force."[70] Negative public opinion was key to settling this dispute, and forced colliery operators to tread carefully in the future over the issue of Asian labour.

The restriction of the Chinese to unskilled jobs was enshrined in law with the passing of the Coal Mines Regulation Act in 1877, which prohibited the Chinese from working in positions of "trust or responsibility." During a strike at Wellington in 1883, however, Dunsmuir hired Chinese strikebreakers to mine coal.[71] The mining population, led by the newly arrived Knights of Labor, now demanded the total exclusion of the Chinese. Their demands received widespread attention and support in this highly racialized province. Businessmen, professionals, and merchants, who "have seen large sums paid away in wages to a class who never enter their stores,"[72] as well as labourers, rallied to the anti-Chinese cause. Politicians were also eager to exploit this issue and threw their weight behind exclusionism.

The government did not respond favourably to the miners' demands for total exclusion in 1883, but in 1884 and again in 1885 it attempted to restrict all Chinese immigration to British Columbia. The province's anti-Chinese immigration legislation was struck down by the federal government, but Ottawa eventually succumbed to provincial pressure and instituted the notorious $50 "head tax" in order to limit immigration. Although public opinion supported exclusion, as long as the two main operators endorsed the employment of Chinese labour, the miners and the provincial government could not successfully exclude the Chinese. With the arrival of Samuel Robins in 1884, who testified to the 1885 Royal Commission on Chinese Immigration that the Chinese should "leave as gradually as they have come into the province,"[73] the capitalist class became divided on this important issue. The most important catalysts for exclusion were the disasters that struck the Nanaimo and Wellington coal mines in 1887 and 1888. The horrific 1887 explosion at the VCMLC in Nanaimo, which killed 96 Europeans and 52 Chinese workers, and the Wellington explosion the following year, in which 68 men, including 37 Chinese workers, were killed, led to renewed pressure to exclude the Chinese altogether. This time the miners were successful. Safety figured prominently in the coal miners' renewed anti-Chinese agitation. Many argued that the explosions had been caused by the carelessness of the Chinese, although there was no evidence to prove this. Miners alleged that because of their lack of English, their "unassimilability," and their alien ways, the Chinese created insurmountable safety hazards underground. In the wake of the explosions, both Robert Dunsmuir and Robins agreed under pressure from the miners – they presented a petition with 1,421 signatures – to ban the Chinese from underground.[74] This was followed closely by an amendment to the Coal Mines Regulation Act in 1890, which prohibited the employment of the Chinese underground.

The Dunsmuirs fought tenaciously against all measures that would restrict the employment of Asians in their mines and ultimately their absolute control over all aspects of their mining operations. In one lawsuit after another, James Dunsmuir contested provincial legislation banning the Chinese and was reluctant to give in to public opinion. Moreover, he achieved his ends, as time and again the federal government struck down exclusionist provincial legislation as unconstitutional. Dunsmuir's victory was far from complete, however. He remained isolated on this issue, and over time the miners succeeded in limiting the employment of the Chinese.

By the time Ladysmith was founded, the only colliery that made extensive use of Chinese miners and labour was Dunsmuir's Union works. When Wellington was abandoned and production began at Extension, Dunsmuir initially "experimented" with Chinese labourers working underground as pushers and tracklayers. However, during the 1900 election campaign, in which James Dunsmuir ran as a candidate for South Nanaimo, he suddenly became an ardent exclusionist. As the Nanaimo newspaper reported, "he had come to look upon their employment as detrimental to the interests of the country."[75] Political opportunism aside, Dunsmuir won his riding by a small margin against John Radcliffe, a labour candidate. When he was called upon to become premier, he honoured his campaign promise by dismissing 100 Chinese labourers who had been working below ground at Extension.

While the provincial press and the Miners' and Mine Labourers' Protective Association (MMLPA) praised Dunsmuir for taking this decisive step towards total exclusion, Francis Little, Dunsmuir's general manager, believed his boss had made a mistake. "He appeared to think he could run as cheaply without them as with them; not a very good result financially," Little believed. "The expense was increased. It cost nearly double in track laying, pushing and that class of labour generally. I have failed to find a single white man that will do the work of two Chinamen at this class of work, and some Chinamen will do that work as much as white men ... I would not change. Mr Dunsmuir wants to change. I do not agree with this new idea of his."[76] In fact, it proved exceedingly difficult and expensive to replace the Chinese. Dunsmuir spent $15,000 to recruit 200 Scottish miners "to enable me to dispense with Chinese labor,"[77] but by early 1901 most had left the Island, in part because of the conditions at the Dunsmuir collieries, in part because of the Chinese, for as Andrew Bryden claimed, "an Englishman or a Scotchman would not take the place of the

Chinese; miners would not."[78] As a result, Little complained: "We have twenty left ... I do not think one-third of them ever dug coal in their life ... I paid $3 a day for a $1 day's work to some of them. I was longing for the Chinamen."[79]

Little's longings went unfulfilled. The number of Chinese labourers employed at Extension was now drastically reduced. Official returns indicate that in 1901, 105 Chinese men were employed above ground but none underground. Dunsmuir refused to remove the Chinese miners from Union. Of a total underground workforce of 627, there were 102 Chinese and 28 Japanese workers. A further 126 Chinese and 34 Japanese men worked above ground.[80] However, their numbers steadily declined over the course of the first decade of the twentieth century. Given this fact, it is difficult to understand why anti-Asian agitation did not also diminish. After all, the Chinese were hardly the economic threat once imagined. Belshaw suggests that hostility towards the Chinese must be seen within the broader context of the erosion of skill and workplace authority and autonomy. In short, the employment of Chinese labour hastened the proletarianization of skilled white colliers. "The de-skilling process that followed the collapse of the strike [of 1883]," he argues, "was so effective that by the 1890s management at Wellington could almost consider white miners dispensable ... In the 1890s the skilled, white Wellington miners were replaceable, whereas the Nanaimo colliers were firmly entrenched."[81]

Apart from the fact that Dunsmuir removed Chinese workers from below the surface at Wellington after 1890, it is not at all clear how individual Chinese labourers eroded the skills of the European practical miners. White colliers did not lose their skills upon the arrival of the Chinese; under normal conditions they did not lose their jobs either, since the vast majority of Chinese workers at Wellington after 1890 were employed at unskilled jobs on the surface. More significantly, skilled colliers were never replaceable. Mining operations on Vancouver Island were generally plagued by a chronic shortage of skilled labour, especially during times of expansion, hence the need to employ inexperienced men. While society at large may not have appreciated the skills of the practical miner, company management certainly did. There is no evidence to sustain Belshaw's categorical claim that the employment of the Chinese was "management's unambiguous statement of contempt for what the émigré British miners valued most: their hard-earned skills."[82] On the contrary, mining officials continually complained about the lack of skilled miners and were always concerned that a worsening of work conditions might drive the skilled away.[83] Similarly, according to the 1902 investigation into the cause of explosions, led by Tully Boyce and John Bryden, the province was "handicapped, owing to the scarcity of skilled miners."[84] A few years later, the inspector of mines made a similar comment, stating that "the general class of mining labour available cannot be selected so as to insure the maximum degree of efficiency, experience, and safety."[85] While the miner's skill may never have completely protected him from replacement by inexperienced men, who, like the Chinese, could learn the skills over time, skilled miners were at a premium in the expanding industry. This, rather than deskilling, may help explain the miners' determination to unionize and become involved in the political process.

Likewise, it is not at all clear that the presence of the Chinese threatened the standard of living of white miners and their families. With the introduction of competition into the labour market, the white miners' ability to restrict access to employment and to determine wage levels was certainly reduced, although it is important not to exaggerate the extent to which workers could influence these factors in the first place. Employing the Chinese for lower rates of pay certainly exerted some downward pressure on wages, especially for unskilled workers. This was especially true during the 1870s and 1880s, when the number of Chinese workers was highest and when wages remained stagnant.[86] At the same time, however, the practical miners who employed the poorly paid Chinese helpers may have actually seen their monthly pay packet improve. Miners also complained that the Chinese took jobs away from their sons and, because they often did domestic service, from their wives, hence reducing the total household income.[87] But was the relatively small number of boys working at the mines due to provincial legislation restricting their employment and to the availability of other employment options or obligations, such as schooling? Was it due to the ready pool of cheap Chinese labour? Similarly, did white women actively seek work as domestic servants? John Bryden contradicted the miners' claims in testimony given to the 1885 Royal Commission on Chinese Immigration, when he suggested that since "few domestic servants come from Europe or America to this colony ... persons with capital might hesitate to come if they knew that servants were not to be had." Historian Patricia Roy also suggests that, generally speaking, "white British Columbians themselves did not want domestic work."[88]

As with the process of mechanization and rationalization, the employment of Asian workers threatened in theory to proletarianize the practical miners and revolutionize the labour process – and hence create an unbridgeable gulf between skilled and unskilled colliery employees – but in practice the social position of practical miners remained relatively stable.[89] Neither mining technique, nor mechanization, nor the employment of unskilled Chinese labourers reduced the need for skilled hewers, who were always in short supply. While the technique of longwall mining had the potential to reduce the miner's autonomy and increase supervision, it neither reduced the status of the skilled miner nor destroyed his hard-earned skills and knowledge. Timbers still had to be set, undercuts had to be made, shots needed to be fired, ventilation had to be sound, gas had to be detected and cleared, and miners had to be able to read the seam and listen to the roof.

The skilled white collier consequently appears to have had little to fear from the employment of the Chinese and was able to maintain his social status and economic position. When all is said and done, the miners objected to the Chinese not because they posed an economic or social threat but because of cultural and, to a lesser extent, biological racism, both of which were pervasive in nineteenth- and twentieth-century attitudes and thought. Although there were many rational economic arguments against the use of exploited Chinese labour to keep wages low, the language used to describe the Chinese was frequently cloaked in racial, gendered, and sexual, not just economic, terms; the miners did "invoke notions of racial superiority and exclusiveness to explain their actions."[90]

Rather than being seen as victims of an oppressive capitalist class who paid them an exploitative wage, the Chinese were considered "servile" and "docile." Rather than include them in the struggle against the companies, the white miners sought instead to exclude them on the grounds that they were an inferior, alien race; they were stupid, obsequious, and posed a serious threat to "public health" and morality because they were sexually deviant and unclean. John Tindal, a Victoria resident, seemed to capture the sentiment of a broad sector of public opinion when he claimed that the Chinese were monsters without morals. Venereal disease, he had been told, was common among them; they had no respect for women – "they seem to think more of a prostitute than they do of a virtuous woman" – and it was a "well-known fact that their women are sold all over the country as prostitutes."[91] According to the Knights of Labor, they were "parasites preying upon our resources and draining the country of the natural wealth which should go to enrich it." "They live, generally, in wretched hovels, dark, ill-ventilated, filthy, and unwholesome, and crowded together in such numbers as must utterly preclude all ideas of comfort, morality, or even decency, while from the total absence of all sanitary arrangements, their quarters are an abomination to the eyes and nostrils and a constant source of damage to the health and life of the community." As a result, they were unsuitable immigrants; they were parasites who undermined and impoverished the nation, were an undignified source of competition to the racially superior white men, and consequently eroded the privileges that were the birthright of British subjects. They were the very opposite of the "free, manly, intelligent and contented laboring population [that] is the foundation and source of the prosperity of any and every nation."[92]

It is often suggested that the employment of Chinese workers "fractured the labour force and helped to frustrate efforts to organize trade unions on the Pacific coast," and that this was the explicit intention of operators.[93] But it was not the Chinese who fractured the labour force, drove down wages, or prevented unionization. White miners knew that by including Asian workers in their labour organizations their position would be strengthened. On numerous earlier occasions, Chinese mine workers expressed solidarity with their European counterparts. Gallacher points out that "circumstances soon drew the Chinese and whites together as parts of a single working group so that by the early 1880s whenever the miners chose to strike over wages the Orientals (along with non-mining whites) invariably followed suit."[94] Such moments of solidarity were fleeting. During the 1903 strike, the Western Federation of Miners (WFM) briefly considered the possibility of organizing the Chinese at Union, but this plan never got off the ground. As organizer Thomas Shenton claimed, because the Chinese were "aliens to a large extent, and they are a people whom the white people cannot compete with ... to endorse the idea of their being put on a level to some extent with us would be detrimental."[95]

In many respects class inclusion was beside the point. European miners knew that their cause would be enhanced, but racial prejudice was stronger than economic common sense or ideological principles.[96] Caught between being the exploiter or the exploited, European miners chose to be the former rather than cease being the latter. Inclusion would have defeated the purpose of the white miners, which was total exclusion, not acceptance, even though this meant they were unable to present a united front. Indeed, the ranks of the white miners would not have been as strong had the Chinese been accepted and included, because the Chinese issue was a central rallying point for the white community.[97] The miners successfully exploited the racial issue for their own political purposes; race empowered and mobilized them politically and economically, all the while inflating their own self-worth. In short, rather than undermining the status of white miners, which in the eyes of the general public was already low, the presence of the Chinese redefined and enhanced their status. Because racism is as much about defining the Self as about defining the Other, white miners were able to forge their own collective identity and "imagined community" by creating a racial Other in the Chinese and by articulating their own "ontological space" through exclusion and isolation. This reinforced their status as members of a socially and racially dominant elite group, as the miners, many of whom were newly arrived immigrants, sought to redefine their identity as citizens of the young Canadian nation.[98]

Work conditions, mining technique, the organization of labour in and about the mines, and racial prejudice shaped the identity and consciousness of European miners and mine labourers in significant ways. In recent years, historians have emphasized how the workplace above and below ground was radically transformed by the introduction (on a global level) of new mining methods, technological advances, and the influx of unskilled, immigrant labour as the coal-mining industry expanded rapidly during the last quarter of the nineteenth century. These developments, so the argument goes, posed a serious threat to the status of the "independent collier" because they destroyed the miner's traditional craft status, eroded his skill, and threatened to reduce him to the level of factory worker. The result was the emergence of a strong class consciousness – what some have labelled militancy and radicalism – that found expression in specific working-class politics and the growth of unionization.

Rather than indicating that the emergence of a strong consciousness of class and identity among colliers was a reaction to some external, structural threat, the example of Vancouver Island suggests that the source of this consciousness emerged from within: from traditional, historical conditions and social and racial relations, and from the complex structure of the workplace and the community. The extent to which the coal-mining workforce was proletarianized through a conscious processes of deskilling and mechanization, or by the employment of cheap Asian labour, is not at all self-evident. What becomes clear is that despite rapid technological changes on the surface, the pace of restructuring and rationalization underground was extremely slow. The introduction of blasting powder, for instance, or improved safety lamps, enabled mines to be dug more deeply and production to increase, but the actual means of winning and loading the coal – by pick and shovel – remained the same, and the stall remained a place where traditional skill and experience still counted. Indeed, the persistence of the "old regime" in mining, rather than its destruction by new management and production techniques or because of the influx of unskilled labour, was a source of strength and power for the miners in their struggles with operators. Not only did it reinforce their traditional social and racial values and sense of community but it was a clear sign that they were not dispensable.

# 4
# *Death's Cold Hand*

"In Memoriam"

Fame awards a crown of valour
To the Men on land and seas,
Who defend our country's honour.
There are men more brave than these

Down in the dark and gloomy dungeons,
Dangers threatening everywhere,
Death's cold hand forever over them,
Yet for self, they have no care.

Only for the sake of loved ones
Does each fearless, noble soul
Go midst many untold dangers,
Go to work amidst the coal.

Hard it is for us to fathom,
In this dark, heartbreaking hour,
Why Death plucked from some home gardens
Just the best-loved favourite flower.

Mothers, wives, and orphans weeping,
Grieving for those hearts so true,
In your hour of deep affliction
Only God can comfort you.

And when Death, the ruthless reaper,
Comes to claim you for his own,
May you find your dear ones waiting
For you by the great white throne.

– Dollie Scannell, Extension, 6 October 1909,
in *Ladysmith Chronicle,* 9 October 1909

Shortly after 8:30 on the morning of 5 October 1909, an explosion ripped through Extension's No. 2 mine, killing thirty-two men. Although the final reports of the inquest were conflicting and inconclusive, it seems that the disaster was the result of a faulty shot fired by James (Juraj) Keserich in room 27, 2½ west level.[1] The blown-out shot triggered a massive 240-foot-long cave-in, releasing a large quantity of methane gas that was then ignited by the open-flame lamp of one of the miners working at the foot of rooms 20, 21, and 22. Sixteen men were killed in this district, including eighteen-year-old mule driver William Quinn. Quinn had been bringing empty coal cars to the face when the blast threw him back into a car. He and his mule were killed instantly. A couple of hundred feet down the level, rescuers found the bodies of fifteen other miners, many of whom were still at their workplaces. Thirty-nine-year-old John Wargo (Ivan Vargo), father of nine and a well-respected, experienced miner who had lost an eye in an accident in 1902, was found at the top of room 21. Apparently alerted by the initial cave-in, he and some other men had left their stalls, only to be killed as the explosion expended itself in their working area, which formed a dead-end or cul-de-sac.[2]

The explosion was localized, but it was powerful enough to produce a second large cave-in and to force the firedamp back along the air course of 2½ level, where it either caused another explosion when it came into contact with James Keserich's lamp or continued its explosion down towards the face of 2½ west level and No. 3 level. Either way, Keserich was in the direct path of the explosion. His body was found just outside his stall and had suffered severe burns to the head and torso.[3] The men in the west level, past room 27 and the initial cave-in and through to No. 3 west level and No. 4 west level, had some warning of the disaster. James Keserich's brother William and his partner, George Bodovinac, both miners, managed to run about 160 feet from the face to the foot of room 25 before being overcome by afterdamp, as did James Molyneux and Thomas O'Connell.

Only the bodies of two men in this district showed significant signs of force. Harold Taylor and Alexander Milanich, known as Milos, were caught in the second explosion at the top of room 25. Taylor suffered severe burns to his upper body. The exact circumstances surrounding the death of Milos perplexed the investigators. One of his boots was found almost forty feet from his body, and he had suffered traumatic head injuries. One investigator suggested that at the time of the blast his foot had become caught in a "track-frog," and he had been blown out of his boot. Another investigator thought it more likely that "in his terror he wrenched his foot free from the boot, tearing the boot and breaking a small bone in his leg in so doing, and in the darkness ran headfirst into a timber, crushing his skull."[4]

The carnage continued further down the level. Two seventeen-year-old mule drivers, William Davidson and Edward Dunn, and the miner Thomas Thomas, who was thought by one of the investigators to have fired the blown shot that triggered the catastrophe, were killed. Five other bodies were also found huddled together at the foot of room 9, No. 4 west level. In their panic to escape the underground inferno, these men had stumbled into a dead-end and were soon overcome by afterdamp. Twelve others, who initially had been with the five men, were luckier. They had decided to take a different route to safety and had made their way to the bottom of room 13, inside the curtain that blocked off the level; they were found alive by the rescue team.

The October explosion, the worst in the history of the Extension mines, was a devastating blow to the young community of Ladysmith. Thirty-two men lost their lives, thirteen women were widowed, and thirty-eight children, only two of whom were of an age to provide for themselves, lost their fathers.[5] It was a solemn reminder not just of the fragility of human life but of the great cost that was being paid by workmen and their families for the town's prosperity and the economic development of the province. It was also a reminder, if one was ever necessary, of the importance of mine safety. In his study of the Vancouver Island collieries in the nineteenth century, Daniel Gallacher argues that safety was never the chief concern of miners and that it was "not a significant issue between labour and management." Instead, the actions of miners and management alike were governed by "a fatalistic view of coal mining, expecting accidents – some disastrous – to happen often."[6]

While the existing structure of economic relations dictated that productivity rather than safety was often the primary concern of miners and operators, it would be a mistake to assume that workers were passive in the face of the daily hazards they confronted. They did struggle for improved safety and working conditions despite encountering significant obstacles, most notably from the operators and the provincial government. This can be best seen in their long attempt to unionize and in their demands for effective state regulation of the coal industry and safety

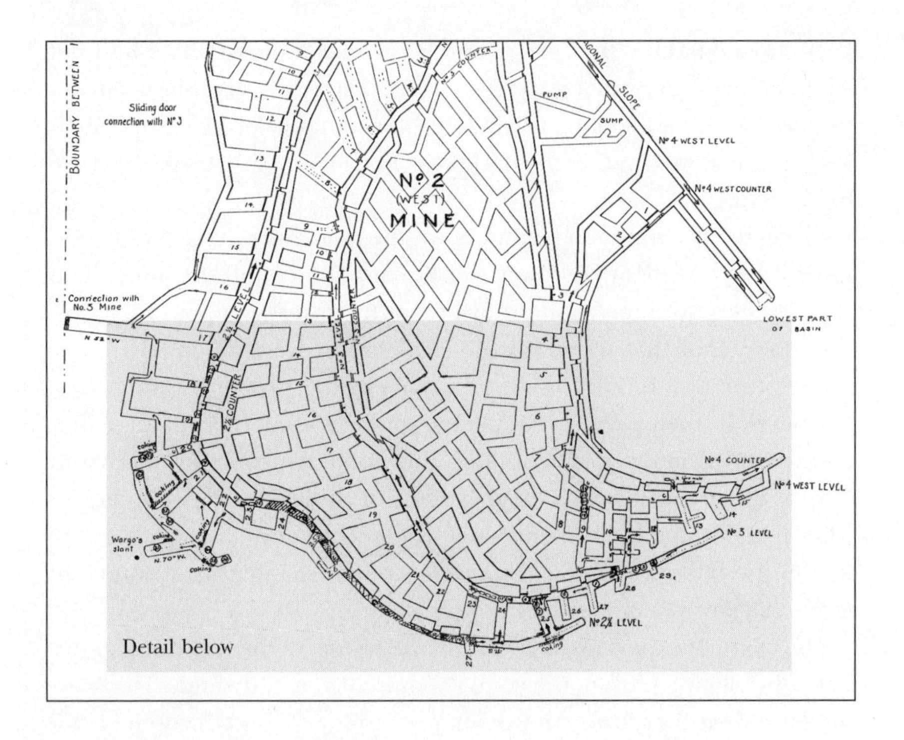


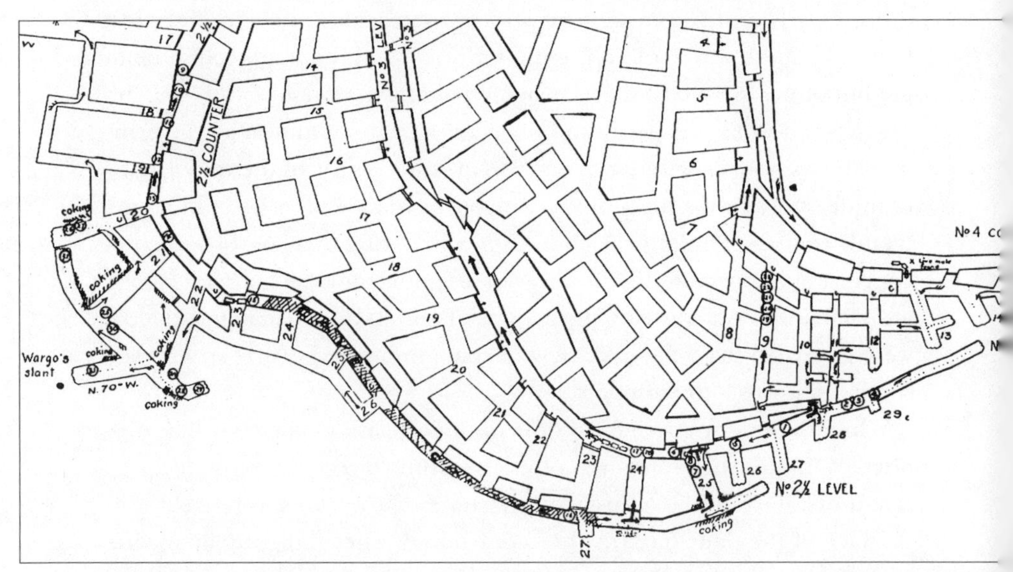


Diagrams of No. 2 mine from the investigation into the 1909 explosion at Extension. The cross-hatched area at the bottom of the diagrams shows the extent of the cave-in. Circled numbers indicate the location of bodies and correspond to the list of men killed (opposite page). From *Annual Report of the Minister of Mines*, 1910.

**List of persons killed in the Extension explosion**

| *No.* | *Name* | *Occupation* | *Age* | *Nationality* | *Doctor's remarks* |
|---|---|---|---|---|---|
| 1 | Andrew Moffatt | Bratticeman | 42 | Scotch | No burns, CO. |
| 2 | Thomas O'Connell | Miner | 27 | Irish | No burns, CO. |
| 3 | Jas. Molyneux | Miner | 36 | English | No burns, CO. |
| 4 | Thomas Thomas | Miner | 29 | Welsh | Slightly burned on back of hands. |
| 5 | Alex. Milanich (Milos) | Labourer | 28 | | Fractured skull and right leg. No burns. |
| 6 | Harold Taylor | Labourer | 30 | English | Severely burned on face, neck, and hands. |
| 7 | William Keserich | Miner | 38 | Slavonian | No burns, CO. |
| 8 | Geo. Bodovinac | Miner | 25 | Austrian | No burns, CO. |
| 9 | Robert Marshall | Pusher | 25 | Scotch | Slight burns on face, chest, and hands. |
| 10 | Wm. Robinson | Miner | 35 | Scotch | Slight burns on face, chest, and hands. |
| 11 | Peter Nieland | Pusher | 39 | English | Deep cut on back of hand. |
| 12 | John Ewart | Pusher | 38 | | Slightly burned on face. |
| 13 | Charles Scheff | Labourer | 27 | Scandinavian | Slightly burned on face. |
| 14 | W.A. Selburn | Labourer | 45 | American | Burned on face, neck, and hands. |
| 15 | William Quinn | Driver | 18 | Canadian | Face burned, gloves on hands. |
| 16 | Alex. Keserich | Miner | 21 | Austrian | Severely burned on face, neck, and left side, chest, and hands. |
| 17 | William Davidson | Driver | 17 | | Slightly burned on face and hands. |
| 18 | Edward Dunn | Driver | 17 | Canadian | Slightly burned on face and hands. |
| 19 | Fred Ingham | Bratticeman | 23 | American | No burns, CO. |
| 20 | Winyard Steele | Pusher | 25 | American | No burns, CO. |
| 21 | Alex. McClellan | Driver | 36 | Scotch | No burns, CO. |
| 22 | Robert White | Miner | 46 | Scotch | No burns, CO. |
| 23 | John Isbister | Tracklayer | 48 | Canadian | No burns, CO. |
| 24 | John Bulish | Miner | 34 | Austrian | Slightly burned on face and hands. |
| 25 | Mike Gustav | Loader | 30 | Austrian | Burns on back of neck and hands. |
| 26 | Oscar Nyman | Pusher | 28 | | No burns, CO. |
| 27 | Charles Salo | Loader | 30 | Hungarian | No burns, CO. |
| 28 | Thompson Parkin | Miner | 27 | American | No burns, CO. |
| 29 | Todd Rombovia | Miner | 42 | Austrian | No burns, CO. (1st degree burns). |
| 30 | John Wargo | Miner | 43 | Slavonian | Burned on face, neck, and hands. |
| 31 | Mike Danculovich | Runner | 25 | Austrian | Burned on neck, face, and hands. |
| 32 | Herman Petersen | Miner | 43 | Dane | No burns, CO. |

legislation. While this struggle may not have radicalized the workforce because the miners' efforts remained within the established economic and political framework, in the end it did forge strong communal bonds that facilitated the political and economic mobilization of the coal-mining population on Vancouver Island.

Death and injury were a fact of life in all coal mines. In almost every aspect of the mining operation, the miner was confronted with life-threatening situations in his daily work routine. Accidents seemed to be random, occurring suddenly and unexpectedly. They struck down the skilled and the unskilled, the man of years of experience and the raw apprentice, the careful and the careless. They could happen in spite of the best regulations and precautions, and often seemed to defy all rational explanation, no doubt contributing to the miner's much discussed fatalism and superstition.

During the late nineteenth and early twentieth centuries, the coal mines of Vancouver Island were among the most dangerous in the industrialized world. In the ten-year period between 1894 and 1903, for instance, 347 men lost their lives, for an average annual death rate of 8.96 per thousand men employed. Between 1901 and 1910, the statistics improved slightly to 7.56 per thousand.[7] Contemporaries were concerned about the high death rates, although perhaps they took solace at the apparently low accident rates. According to official figures, the rate of injury (both serious and slight) between 1892 and 1903 in British Columbia mines was 14.17 per thousand men employed, a rate considerably lower than in European mines; in Britain between 1885 and 1897, the average nonfatal injury rate was 123 per thousand men employed.[8]

Explosions were the most dramatic and catastrophic cause of death in the Island mines. Between 1894 and 1903, 63 percent of all fatalities, or 219 men, were caused by explosions.[9] The main cause of explosions was the ignition of methane gas ($CH_4$), known by miners as firedamp. Methane, a natural product of the decomposition of the organic matter that forms coal, is present in all coal mines and its introduction to the air is a natural consequence of mining. Because coal is porous, it absorbs the gas. The process of mining, however, creates fissures and fractures that allow methane to migrate, seep, and bubble to the exposed coal surfaces, where it is then released into the atmosphere underground. When present in sufficient quantity, methane is highly flammable. We now know that the explosive zone of methane is between 5 and 15 percent, and the slightest spark – "as little as ... about one fiftieth of the static energy accumulated by an average-sized man walking on a carpeted floor on a dry day" – not to mention a miner's open-flame lamp, can cause the gas to ignite and explode.[10]

While a methane fire or explosion could take a horrific toll in lives, not all victims succumbed in the initial blast. The combustion of the burning mixture of methane and oxygen produces another gas, carbon monoxide, known by miners as afterdamp, which kills by suffocation. Often, as at Extension, men would survive an explosion only to be overcome by afterdamp as they stumbled to escape through the wreckage and dark tunnels underground. The extent of the explosion also depended upon the amount of fuel available to feed the fire. Local methane explosions were often propagated throughout extensive areas of a mine by the presence of high levels of coal dust. Coal dust results from the constant drilling and cutting of the coal and does not require the presence of gas to explode; it is explosive by itself. All that is required for a coal dust explosion, according to one recent study, is "a preceding event that (1) generates a shock wave capable of raising settled dust into the air in high concentration, and (2) produces a temperature high enough to ignite the dust. These two conditions are fulfilled by a methane gas explosion."[11]

Mining engineers and miners did not recognize the danger posed by coal dust until the 1870s and 1880s, and even then many still believed that without the presence of firedamp, coal dust at worst could cause only local explosions over short distances. On Vancouver Island, some experts were aware of the danger posed by coal dust. John Bryden testified that he feared coal dust more than he feared gas but was reluctant to water down the coal because of the threat of spontaneous combustion in mines containing iron pyrites. He also believed, like many others, that open-flame lamps posed no serious danger. The Nanaimo explosion of 1887, which took 148 lives, changed some people's minds. Coal dust had certainly been a factor in the propagation of the explosion, and immediately afterwards extensive networks of underground pipes were installed to spray dry areas at the Nanaimo and Wellington mines.[12] Not everyone was convinced about the danger posed by coal dust, however. As one local newspaper reported in the wake of the 1909 explosion, "there is still a great deal of scepticism as to coal dust acting as a chief factor in an explosion."[13]

Coal dust could be neutralized by spraying water or spreading inert, noncombustible stonedust such as limestone or dolomite, but the best way of reducing the buildup of methane gas was to install effective ventilation systems. This was especially necessary at the exposed surfaces of the face. Over the course of the nineteenth century, significant improvements were made in mine ventilation. Traditionally, mines had been ventilated by furnaces underground. A furnace at the bottom of a shaft acted as a chimney, forcing stale air out of the mine while at the same time drawing fresh air into the workings through a separate air shaft. The installation of steam-powered and ultimately electric fans significantly improved underground ventilation, although they did not completely eliminate the problem of providing sufficient air for the miners underground. This was especially true in pillar and stall mines, which required complex courseways to direct the flow of air throughout the maze of tunnels, crossways, and stalls.

Even with high-powered fans and a sophisticated network of doors and brattices to direct the current of air, it was often difficult to maintain an adequate airflow. In addition, because methane is lighter than air, its buoyancy causes it to accumulate in cavities and crevices that might not be diluted by the stream of passing air or detected by testing.[14] During the inquest into the Extension disaster of 1909, important questions were raised about the quality of the ventilation. While the majority opinion argued that the flow of air was sufficient to dilute the methane gas, James Ashworth, who had been engaged by the provincial government to investigate the explosion, argued that the amount of methane gas, at least 3 percent in the air, had at a minimum rendered it unsafe to fire powder shots.[15]

Underground fires were another common hazard in the Island mines. Although they did not take as high a toll of lives as explosions, death in such cases was just as gruesome. One of the worst fires in the history of the Island mines occurred in the No. 2 mine at Extension in 1901. The fire was started by the ignition of a curtain and killed sixteen men, most of whom died from the dense smoke that filled the tunnels. Fires often resulted from carelessness or explosions, and because of the almost limitless fuel underground, they were exceedingly difficult to extinguish. The Extension fire began on 30 September 1901 and burned until late February 1902, when it was decided to flood the mine. Accordingly, a dam was built to block the tunnel and water was pumped in. On 24 February, mine officials determined that the fire had been extinguished. It took a further month to drain the flooded area and even longer to repair the damage underground.[16] Sometimes flooding did not extinguish an underground fire. One fire in a Pennsylvania anthracite mine began in 1859 and burned for eighty years.[17]

Falls of rock and coal were a far more common cause of death than fire, and second only to explosions on Vancouver Island. They accounted for the greatest number of casualties in American and European mines, and were more frequent the deeper the mine because of the increased pressure on the coal from the overlying strata. Because they usually occurred singly – one historian has referred to them as "colliery disaster in instalments"[18] – they were not as dramatic and catastrophic as explosions, but they were frequent and sudden. Between 1894 and 1903, sixty-four men were crushed to death by falling rock or coal. The risk of falling rock or coal was particularly acute after the firing of a shot to bring down the coal, because the blast often weakened the roof or knocked down the props.[19]

Such accidents not only took a heavy toll of lives but also caused innumerable injuries. The list of men who suffered broken arms, legs, ribs, and skulls was quite extensive. Between 1894 and 1903, falls seriously injured 141 miners and slightly injured another 36. While slight injuries could result in the loss of one or more days of work and the cost of medical expenses, serious accidents could permanently disable a miner, ending his working career and leading his family to destitution. Other men were burned or disfigured in minor gas explosions or when lighting their shots. In 1902 the Ladysmith miner John Wargo lost an eye in a premature blast when he lit the fuse before putting powder into the hole. He survived only to lose his life seven years later in the 1909 explosion.[20] A few months after Wargo's injury, his fellow worker John Barlow had an eye blown out and suffered serious facial injuries in a delayed-shot explosion.[21] Others, such as the Chinese pithead worker Mow, were killed when they fell into open shafts. A similar fate befell L.C. McDonald, a carpenter at Union, on 21 July 1900. In the words of the inspector of mines, McDonald "was instantly killed by falling down No. 6 shaft, a distance of 360 feet. Deceased and two other men were on the cage at the time, repairing the air-shaft, and in knocking off a batten a little gas burst out from behind the timber putting out their lights. Deceased must have become excited and fallen off the edge of the cage."[22]

More frequent than excited men falling down shafts were accidents and fatalities involving the coal cars. The transportation network above and below ground was very extensive and dangerous. Derailments and runaway coal cars were common, and the narrow tunnels often did not have enough room for both a passing train and a passing man, although the Mining Act required "man-holes" every twenty yards for this very purpose. In the ten-year period from 1894 to 1903, twenty-five men were killed by coal cars and a further ninety-one were injured. Often runaway cars killed men who were caught in their path; some were dragged or squeezed between the wall of the tunnel and the box, others lost limbs. Despite being careful, George Cripps, a mule driver, was unexpectedly crushed to death between full and empty cars when his mule suddenly bolted. One unidentified Chinese labourer at Extension was killed at the pithead when he slipped in front of a loaded train of cars. Joseph Crosette, a Union miner, survived being run over by a trip of cars, but had to have his leg amputated shortly before Christmas of 1903. Crosette was fortunate. In 1904 James Strang, an overman at Union, died after his leg was severed by a "coal car in motion."[23]

The geological or physical structure of the coalfield had a great impact on levels of danger and safety, as well as on mining operations generally. Contemporary mineralogists and geologists were unanimous in describing the Vancouver Island coalfields as geologically unstable.[24] According to one colourful and enthusiastic writer, Vancouver Island's "coal-bearing strata were raised from the 'vasty deep' and tilted up, at various angles, so that after the glacial erosion, the rocky escarpments were left in view with seams of coal exposed, and these when discovered were mined from the daylight, down, until Dame Nature's obstacles cut off pursuit. And thereby hangs a tale, for just as in human life the good is mixed with ill, so the first beneficent rising of the giant mountain being all good, his after shaking and lateral push were evil of the direst kind, and many a 'fault', 'synclinal' and 'anti-clinal' crush, with 'troughs' and 'pockets' large and small, and total 'wants', and troubles innumerable, are the coal operator's opprobia and inheritance, and the Island coal seams have their full share of these incommoding and expensive ills."[25]

In fact, "Dame Nature" was more than a little incommoding. Throughout their existence, coal-mining operations on the Island were plagued by the geological irregularity of the three main seams of coal, the Wellington, Newcastle, and Douglas. While all three were faulty, the Wellington seam, which extended from Extension north to Comox and was mined by the Dunsmuirs, was most notorious for its folding and overlapping faults and other distortions in the strata.

These irregularities in the roof and floor of the seam produced "pinches," the narrowing of the coal deposit, and "rolls," horizontal breaks in the coal deposit caused by severe changes in the dip or inclination of the strata. The presence of rolls and pinches had a number of important consequences for mining operations. First, the seams were highly irregular; the thickness of the Wellington seam, for instance, varied "from almost nothing to over thirty feet, sometimes within a lateral distance of less than 100 feet."[26] Second, this greatly affected the quality of the coal. Due to the pinches and rolls, the seams varied from good-quality bituminous coal to poor-quality "rash," or coal with over 50 percent ash content, all of which were interspersed with many barren places or "wants," containing rock, shale, and other waste. This made the coal difficult and expensive to extract. The mines produced an enormous amount of waste material, which the miner had to dig and then separate from the clean coal below ground, and required a great deal of "narrow" work, all of which made it more difficult to earn a living wage.[27]

Many experienced miners grumbled about the irregularities of the seam, especially those newly arrived from Britain, where the seams were thought to be more uniform.[28] They also complained bitterly about being forced to work in "deficient" places, that is, places where coal was either difficult to extract or where there was too much dirt or rock. If the place was deficient, the miners were sometimes given extra pay in compensation. The difficulty, as company officials did not hesitate to point out, was distinguishing between a deficient place and a deficient miner.[29] Some men even refused to work the awkward Island seams altogether. One noteworthy occurrence was in 1900, shortly after James Dunsmuir decided to remove the Chinese from below ground in Extension. To replace the Chinese, he imported about 200 miners, mostly from Scotland. After paying their fare and dividing them up between the Union and Extension mines, he found that most of them refused to work and left for Washington state and the Kootenays. Although some argued that the company had not lived up to its contract with the men, which was highly likely, Andrew Bryden, manager at Extension, maintained that "the work here was different and they did not like it ... They were an average good lot of miners and after two or three months' work here would have been efficient for our work."[30]

The irregular nature of the seams also made the coal more dangerous to mine. The rolls and pinches in the strata created unstable conditions, making the Island mines especially prone to falls of rock and coal and increasing the danger of gas explosions, since both faults and falls released methane. The danger was intensified if the rolls were not detected beforehand. Initially, an undetected roll was thought to have caused the explosion at Extension's No. 2 mine in 1909. According to the official verdict of the jury of inquest, the explosion was the result of a roll or fractured line in the strata, which collapsed, releasing a quantity of firedamp. This was then mixed with the low percentage of gas already in the atmosphere and subsequently ignited by a naked-flame lamp used by one of the men in the vicinity of the roof fall.

Despite the acknowledged geological peculiarities of the Vancouver Island coalfields, provincial mine inspectors blamed many of the accidents and deaths on miner carelessness. Thus in 1891 we read of a "reckless and venturesome disregard of careful propping of the roof and spragging of the coal." A decade later, the miners' work habits apparently had not visibly improved. "With regard to the above serious and slight accidents," the inspector reported in 1901, "in nearly every instance they occurred through want of thought and care on the part of the workman; and even in the cases of burns from gas, the persons injured have mostly brought the trouble upon themselves, by knocking down canvas and brattice with shots in their working places, and neglecting to report and have the same made secure before returning to their work." Again in 1904 accidents were attributed to "negligence on the part of the sufferers."[31]

There is little doubt that careless practices and errors in judgment sometimes caused accidents, but the tendency to blame most deaths and accidents on the workers' behaviour reveals more about the contradictions and conflicts present in the system of mining operations than it does about actual miner carelessness. On Vancouver Island, as in most nineteenth- and early-twentieth-century coal mines, safety was governed by a division of labour that assigned primary responsibility for the prevention of accidents in the workplace to the individual miner. Management was responsible for "providing safety-related 'public goods,'" such as "ventilation, mine gas inspections, watering of coal dust to prevent the spread of mine fires and explosions, and provision of pre-cut timbers to use as roof props,"[32] while miners had to ensure against those dangers "which are naturally incident upon the work" of mining coal, such as roof falls, explosions of gas, and other hazards. They, not management, had to ensure that their working place was properly timbered, for instance, or that they followed the proper drilling, blasting, and housekeeping procedures. As one report concluded, the miner "does assume the risk of latent as well as patent dangers, which are a natural incident to the service, and which it is not the duty of the master to guard against ... Hence, those dangers which an experienced miner knows must and do threaten him at all times are an incident to the service, and assumed by him."[33] The legal regime of the "doctrine of individual responsibility" and the "voluntary assumption of risk" was also accepted in BC, as mine inspector Thomas Morgan noted in 1902 when he claimed that "in their working places," the men "were themselves responsible for their own safety."[34]

This laissez-faire approach to safety reflected key assumptions about existing capitalist relations and confirmed management's commitment to the principle of the "lower-cost" preventer of accidents; in other words, safety, like production, was governed by the imperative of maximum profit, minimum expenditure.[35] It also placed the miners in an extremely difficult position. On the one hand, miners' control over safety appeared to reinforce their claim of "organic" control over the work process.[36] As skilled and independent producers, they considered themselves to be the best guardians of their safety underground. The notion of individual responsibility could therefore be seen as a confirmation of the miner's status as an independent collier, of his knowledge and expertise, and of his autonomy and authority underground. On the other hand, this approach to accident prevention created a serious conflict of interest, not only between miner and management but more significantly with the miner's need to produce coal. The doctrine of individual responsibility was therefore a double-edged sword: it helped miners preserve their autonomy underground but it seriously undermined workplace safety.

Issues of safety and wage rates were closely connected. The piece-rate system, whereby the miner was paid by the quantity and quality of coal he produced, was the main form of payment in almost all the mines of the industrialized world. On the surface, piece-rates were advantageous for both the miner and the operator. Like the notion of individual responsibility, this system of payment was a measure of the miner's independence, a reflection of both his intimate knowledge of the mining process and his indispensability within the productive system. It placed the onus of productive and earning power squarely on his shoulders by enabling him to earn relatively high hourly wages, to work largely unsupervised in his stall, and to control his own level of work comfort and production. Operators benefited because the piece-rate system was a more or less accurate measure of the miner's efforts, and because they could reduce the cost of supervision.[37]

As miners recognized, however, the piece-rate system was incompatible with the implementation of safety regulations, because it fostered risk taking.[38] Because they were paid according to the amount of coal they dug, time spent doing other jobs, such as timbering or cleaning their workplaces, reduced their income, since they were seldom adequately remunerated for them. As a result, they were often forced to cut corners or gamble with their safety in order to earn an adequate, living wage. This conflict was highlighted by the *Nanaimo Free Press* in the wake of the Extension explosion. "The digger is on contract, and his earnings depend upon the amount of coal he sends over the company's weigh. There is only a limited supply of cars for him to load his coal into, and if the car is not loaded when the pusher calls for it there is a dead loss of say fifty cents. Hence the hustle and hurry, scheming and striving of the digger to keep his turn, and hence the temptation to take the most foolhardy risks. He has a hole ready but must wait for the fire boss. There is an empty car standing, and the fire boss is not in sight. Does he wait?"[39]

If piece-rates represented a trade-off for their autonomy underground, then safety was often a trade-off for earnings.[40] Perhaps the most obvious example of this was the use of safety lamps. While safety lamps clearly minimized the risk of explosions, miners objected to them because they did not produce as good a light as open-flame lamps. This, the men argued, created new dangers, as "accidents from other causes would inevitably be increased" because hazards would be harder to see, and they damaged the eyes.[41] More importantly, reduced visibility meant reduced production. At the Vancouver Coal Mining and Land Company (VCMLC) at the turn of the century, management agreed to pay an allowance of twenty-five cents to those who had to use safety lamps in gassy districts. After the sale of the company in late 1902, the new management of the Western Fuel Company stopped the allowance, leading to a short strike in February 1903.[42] The issue of safety lamps was not effectively resolved.

The special commission that investigated coal-mine explosions hesitated to recommend their mandatory use, and many men continued to take the trade-off, risking their lives and those of their fellow workers.[43]

In and of itself the piece-rate system did not need to heighten risks, provided the miners received adequate compensation for their work. Higher levels of earnings tended to lower the need to take risks; lower wage levels heightened risk. Daily wages in the Island mines, as in the industry in general, were comparatively high – in 1914 miners were paid between 68 cents and $1.10 per ton, depending on the place, and earned on average between $3 and $5 a day. As Belshaw has argued, however, a host of factors, including a high cost of living, reduced opportunities for the employment of boys, inadequate compensation for dead time, the irregularity of work, and numerous deductions and other expenses, not to mention deficient places, shortage of cars, poor ventilation, and so on, meant that it was still difficult to earn a living wage and escape the cycle of poverty and indebtedness that plagued coal-mining communities.[44]

Despite high nominal wage rates, management policies had a direct impact on real wages and hence safety. The Dunsmuirs, for example, were notorious for devising ways to reduce wages. Nonpayment for dead work – work that did not directly produce coal, such as digging dirt or timbering – was one tactic that was common in the Dunsmuirs' mines. They were also accused of using rigged scales, an issue that sparked the first major strike at Wellington in 1877. Some miners had great disdain for the checkweighmen. One Cumberland miner considered them "nincompoops, good-looking idiots and know-nothings that we have to buck against and be barefacedly robbed by."[45] This was a consequence of the notorious practice of docking, whereby the miner had his car confiscated if it contained too much slack. This was known as the "courthouse" system, which was described in a 1914 Royal Commission investigation of coal in the province: "If, at the tipple, a car is found – either by weight or inspection – to contain excess rock, it is diverted and dumped. And in a typical example, the miner is docked in double the weight up to and including 50 lb., his car is confiscated if the weight is over 50 lb. and including 100 lbs., and if over 100 lb. it is in the hands of the mine to dismiss, which would be liable to follow if, after investigation, deliberation is evident."[46] Martin Horriban, a miner at Wellington, had direct experience with docking: "and in the next place I loaded a car of coal, and the weigh boss up there ... stole it from me; wouldn't give me anything for it. I asked him the reason why he took it, and he said I loaded two pieces of boney in it. He showed me the boney, and I asked him where he put the rest of the coal, and he said he dumped it in the shute, and I guess the company sold it. The two pieces of boney, I guess, would weigh about 75 lbs; no more anyway."[47]

Complaints were also heard about expenses associated with mining, such as the need for the men to purchase blasting powder from the company, a source of considerable profit for operators. According to one disgruntled Cumberland miner, the company bought powder for six cents per stick but sold it to miners for fifteen cents, for a tidy profit of around $10,000 per month. These additional costs were of grave concern to miners and often resulted in job action. When a more expensive explosive was introduced at Extension in late 1910, the miners protested. Only after a period of sustained agitation and an interview with the mine superintendent by a committee of men was the issue settled.[48]

The conflict between earnings and safety was most intense during periods of irregular work or rapid expansion, when demand for coal was high. On the Island, the mines operated an average of 222 days per year.[49] Seasonal layoffs tended to happen during the spring and summer months, when demand for coal for heating in Victoria and Vancouver was low.[50] While some miners might have been willing to make an "income-leisure trade-off," that is, were willing to sacrifice some of their earnings for more time off, especially during the summer months, it generally meant an increase in the intensity of work when it was available, and hence higher accident rates.[51] The same was true during times of growth and expansion in the industry. The provincial inspector noted in 1910 that the high number of accidents from mine cars and horses "bears a direct relation to rapidly increased outputs."[52]

This was compounded if the miner was assigned a poor place. The miner was paid by the ton, and if he was assigned a dirty, rocky place or a narrow seam, he might not be able to earn as much money as a miner in a good place. On the Island, management assigned the places and was often accused of favouritism. Those who did not cause trouble were assigned good places, where an able miner could earn up to $6 or $7 a day. According to the testimony of Aaron Barnes, it was Dunsmuir policy to "have a number doing well, in order to keep the balance quiet." Agitators or union men were often assigned deficient places, if they were not dismissed altogether. In many collieries, including Wellington, a minimum wage of $3 a day was established as compensation, but this was not always put into practice.[53] "Some men get paid for deficient work," continued Barnes, "others don't get paid for it at all." Samuel Mottishaw, a union organizer, recalled digging a total of seven tons in one shift, four of which were dirt and three of coal, but being paid only for the coal. When David Jones asked the foreman to compensate him for his deficient place, he was told to leave if he did not like it: "There are lots of men around glad to get a chance to work at it." The result was that earnings could vary extremely from month to month. Jones's deficient place, combined with the irregular number of workdays, meant that in February 1890 he made $30; in March, $65 working every shift the pit was operational; in April, $56; and in May, $27.[54]

One potential solution to this problem was the cavil system, whereby places at the face were assigned to the miners on a quarterly basis by drawing lots. This lottery, traditional practice in the north of England, was favoured by the men because it reduced the possibility of favouritism and created equal opportunities for good and bad places. Management opposed cavilling, however, because it would have reduced its authority and eliminated a powerful means of disciplining the workforce.[55]

The doctrine of individual responsibility placed miners in a catch-22 situation because low wages and the pressures of the piece-rate system meant that safety was often sacrificed for production. Moreover, the division of safety labour enabled management to lay blame for accidents squarely at the feet of the workers, and to avoid the responsibility for creating a safe working environment. Operators often placed profit before safety. The complaints of miners and provincial mine inspectors reveal not only a high degree of concern about safety standards but also frustration at the reluctance of some operators to take necessary precautions, to supervise effectively, and to enforce proper mining procedures. The Dunsmuirs appear to have been especially complacent in this regard, although until the explosion of 1887 at Nanaimo and the arrival of the new manager there, Samuel Robins, the VCMLC did not have a much better record.[56] The most common complaints appear to have been about poor timbering practices and inadequate testing for methane, both of which were of vital importance given the fractured nature of the seams and the frequency of falls.

The overman or fireman was responsible for inspecting the roof, testing for gas, and generally ensuring that the conditions underground were safe. Often he would order work at the face to stop until the stall was properly timbered. This often produced a dilemma for the miners, because there was no specific payment for timbering. At Wellington, the rule of thumb was that miners were paid $1.25 for each set of timbers, but often they received no payment or were forced to pester the boss for the money. The Belgian miner Gustave Denouve, for instance, was ordered to put up three sets and then told that if he wanted to put up more he would not be compensated for them. In the end, he had to install seventeen sets but was paid for only ten, and then only after "hunting the boss every day."[57]

Timbers, of course, were expensive. For this reason, Wellington manager John Bryden changed the practice whereby miners cut their own timbers. He thought it better for the company to cut the props because he "found that when a man required a prop perhaps eight feet long he would find a stick perhaps ten feet long, and take it down the mine and cut two feet off – that was a loss of two feet."[58] In order to keep costs down, the company not only regularly refused to pay for timbering but was often reluctant to allow miners to timber at all. Victor Delcourt testified that he had been working at a place in Wellington No. 5 where the roof was all dirt for about twenty yards. "Every morning when I was going down," he recalled, "I asked them to put stringers in there, but they said they couldn't put timbers down there." Only after the mining inspector visited the place and told him to leave because the roof "might come down at any time" did the company allow Delcourt to timber his place.[59]

Testing for methane was also haphazard. The use of safety lamps was not required in all mines, even though the most technologically advanced ventilation system could not prevent gas from accumulating in the nooks and crannies of the maze of tunnels and slopes, and certainly could not prevent it from being produced from a fall or cave-in. It was the responsibility of management to inspect the mines for gas and, if present, to ensure that it was eliminated from the working surfaces of the mines. This was not always done. The inquiry into an explosion on 17 April 1879 at Wellington, which claimed the lives of eleven men, revealed the Dunsmuirs' cavalier attitude towards the presence of dangerous levels of gas in their mines, and the difficulties the inspector of mines had enforcing existing regulations in the face of a recalcitrant company. Although the inquiry absolved management of all responsibility for the disaster – it was blamed on a Chinese labourer – Edward Prior, then inspector of mines, testified that there was "a good deal of explosive gas being given off in the tenth level," where the explosion occurred, and that he had "many times warned James Dunsmuir of it."[60] Dunsmuir admitted that Prior had "several times [drawn] my attention to the gas in that level," but Dunsmuir felt that the supply of air was adequate to dilute it. The evidence he gave suggested the very opposite. On a number of occasions, gas had exploded in No. 10 level, burning the men. At the time Dunsmuir thought that "the explosions were trifling" and that the "accidents have been slight" – certainly not enough to warrant taking the appropriate precautions.[61]

It was exceedingly difficult to improve underground safety. The situation was indeed paradoxical, as the frustrated mine inspectors knew.[62] A 1903 special commission that investigated the cause of explosions had revealed glaring shortcomings in underground safety and had made a host of recommendations, including the demand for more vigilant enforcement of regulations, better-qualified overmen, firemen, and shotlighters, the use of "the most competent and reliable workmen" in areas "involving extra risk and responsibility," and stricter "discipline ... amongst both officials and workmen."[63] Exhortations and education were not enough, however; the commission's recommendations had to be balanced with the existing system of production and social relations and with the difficulties of enforcement. Given the almost chronic shortage of skilled labour during a time of expansion, for instance, inexperienced workmen had to be hired. Likewise, greater discipline of the workforce, a common exhortation, was also difficult to implement. Operators would have been forced to hire more supervisors, which they were reluctant to do, and risk alienating scarce skilled miners, who resented the erosion of their autonomy and authority underground.[64]

Yet the miners were not indifferent to the dangers they faced, and they increasingly challenged management's laissez-faire approach towards safety. While they acknowledged their responsibility for workplace safety, they also continuously sought to improve conditions underground through organization, agitation for better wages, pressure on management to fulfill its safety obligations, and demands for increased state regulation of the industry, which many believed was key to improving colliery safety and reducing tensions between them and management. Miner agitation was key to the passage of the Coal Mines Regulation Act (CMRA) of 1877, a derivative of British legislation and the first significant attempt by the state to establish statutory guidelines for the operation of coal mines in the province. Passed into law in spite of intense opposition from Robert Dunsmuir and the other colliery operators, who resented all interference in their workings, the CMRA regulated the age of employees, the hours of work, the use of safety lamps, ventilation, shotfiring, and a host of other practices. It also required certification of managers and outlined the responsibilities of the inspector of mines.[65] As the coal-mining industry expanded over the course of the next few decades and accident levels did not improve, important amendments were made to the CMRA in an attempt to improve the safety of the province's mines. Amendments to the act usually followed catastrophic disasters, such as the Extension explosion of 1909 and the earlier disasters at Nanaimo and Wellington in the late 1880s.

Pressure from the working class was also instrumental in the evolution of workmen's compensation legislation, which gradually eroded the notion of individual responsibility.[66] The Employer's Liability Act of 1897 and the Workmen's Compensation Act of 1902 made employers more accountable for injured workers, but blame and culpability were still assigned to workers through the often lengthy litigation of compensation cases in court. The passage of the Workmen's Compensation Act of 1916 and the introduction of no-fault insurance marked a significant change in policy and philosophy, which had far-reaching consequences. While it no longer assigned blame to individual workers, it also released employers from liability for accidents and injuries. Accidents, rather than being a consequence of work conditions and the system of production, over which the employer exercised direct control, were now seen by the state as the legitimate price to pay for the system of industrial capitalism, the costs of which were to be borne by society at large.

Increased state regulation of the industry and the evolution of occupational health and safety legislation represented important victories for the working class. This was also the case with the provincial government's repeated attempts to exclude Asians from working underground and its requirement for certification of miners. Not only did these laws meet the European miners' perceived belief that the Chinese posed a threat to safety but by restricting the employment of inexperienced men, whether Chinese or European, skilled colliers enhanced their status and ability to control access to the trade. This was no doubt a blow to the Dunsmuirs, for whom the Chinese were not just a means of enhancing profits but "a means of loosening the grip of colliers' organizations on the labour supply."[67]

Coal operators resisted government attempts to regulate the industry. The Dunsmuirs fought the legislation at every opportunity, believing it to be an unwarranted instrument of political interference. Ultimately, it threatened their authority in the workplace and empowered the miners. As a result, shortly after the CMRA became law in 1877, James Dunsmuir claimed that "he did not care for or think anything of the Mining Act as it was altogether unconstitutional and he would prove it shortly in a higher court."[68] Indeed, he went to extraordinary lengths to neutralize it, especially with regard to the Chinese exclusion clause.[69] While the CMRA appeared to go against the interests of the most powerful industrialist in the province, the state was still nevertheless clearly reluctant to create legislation that would hamper the profitability of big business or interfere in its managerial prerogatives. Legislation did nothing to alter the existing social or economic relations.

This had a number of consequences, especially with regard to safety. First, the intent of the legislation was not just to create a safer working environment but to create an arena in which tensions between labour and capital could be voluntarily resolved. Just as the state was reluctant to interfere with profits, it was also reluctant to penalize violations of the law. In fact, penalties under the law were negligible. On numerous occasions, as, for instance, when Francis Little, the manager of the Union colliery at Cumberland, was charged in 1890 with violating the CMRA amendment that prohibited the employment of Chinese underground, the magistrate dismissed the charges because there was no penalty for the violation.[70] Enforcement of the laws was therefore extremely problematic. As Tully Boyce stated: "Unless the owners have a dread that the law will be enforced, they generally take chances."[71] Clearly there was little to dread about the CMRA. Over two decades later, the socialist newspaper the *Western Clarion* was still complaining that "the mines were allowed to be run as the mine-owners saw fit." No operator was ever prosecuted for negligence causing death or injury of a worker, the newspaper report continued. Moreover, miners who made claims for compensation for injuries suffered on the job often risked being blacklisted.[72]

Second, the lack of penalties and enforcement meant that the issue of management responsibility for safety was never adequately resolved. Legislators clearly believed, and still commonly do, that management shared the same interest in safety as the miners. Accidents, after all, could seriously hamper production and consequently reduce profits.[73] The evidence suggests, however, that the profit imperative weighed more heavily than safety, and that in the absence of the coercive force of the state, operators continued to put the lives of their employees at risk.

The problem of enforcement of BC's comprehensive mining legislation was further complicated by acknowledged deficiencies in inspections by the provincial department of mines. In the wake of the Fernie disaster of 1902 and the publication of the subsequent Special Commission Appointed to Inquire into the Causes of Explosions in Coal Mines, opposition politicians condemned the lack of enforcement by the department. According to James Hawthornthwaite, "inspection is to a large extent a farce, the inspectors being clearly given to understand that there are many things which they should see that they must not see if they value retention of their offices." Instead of being appointed "by the government of British Columbia [which can] not get away from the domination of colliery interests," inspectors should be "named, not by a class-controlled government, but by the miners whose lives were at stake upon their devotion to their responsibilities."[74] Provincial inspectors, however, responded defensively to such criticism and rejected out of hand any suggestion that inspections were "farcically imperfect" or that their numbers were inadequate. In 1909 James Ashworth rejected the call for more inspectors, fearing that this might lower their status.[75]

Agitation for protective legislation, enforcement of regulations, and improved inspections went hand in hand with attempts to organize. While Fishback and others have argued that unions – in this case the large United Mine Workers of America – had little if any impact on accident rates and that safety was not a primary concern of the union,[76] unions, by empowering the workforce, enabled miners to resolve some of the tensions and conflicts that stood in the way of improved working conditions and safety. This can be seen most clearly at the Nanaimo operations of the VCMLC, where accident and fatality rates improved substantially after 1890 and the unionization of the firm's workforce. Although the geological structure of the seams, the production methods, and the wage system were the same as in the Dunsmuir mines – in fact, the Nanaimo mines were potentially more dangerous because they stretched under the harbour – Nanaimo's safety record was much better. While the data are incomplete, largely because Dunsmuir, as was his legal right, did not always supply official returns to the Ministry of Mines, the information we do have follows a consistent pattern. Thus, in 1900, a year with no serious explosions, the death rate at Nanaimo was 1.61 per thousand men employed. At Extension, the rate was higher at 2.83, while the rate for the Union colliery was 8.85. The following year, due to catastrophic accidents at both Union and Extension, the death rates skyrocketed. It was 33.78 at Extension and a stunning 80.59 at Union, while Nanaimo's death rate was 2.67 per thousand men employed.

Improved labour relations were essential to the improvement of underground safety. The union provided the miners with a mechanism for airing grievances without fear of retribution and a strong incentive for management to follow and enforce regulations. Miners were also more actively responsible for all aspects of mine safety, not just that of their stall, because they could form pit committees, one of the primary objectives of which, according to Robins, was to check "the safety of the mine, both in regard to ventilation and in regard to any method of getting the coal that might be considered unsafe."[77] Pit committees empowered the men because they acknowledged their authority in the workplace and were vital in settling numerous small and large grievances and bad conditions that could disrupt production.[78] While they were a common feature in the Nanaimo works, they were not tolerated at the Dunsmuir mines. As Archibald Dick, inspector of mines, reported in 1896 and again in 1897: "Once more have I to record that the miners of the Nanaimo Colliery are the only workmen who have as yet availed themselves of the privileges allowed them under General Rule 31, 'Coal Mines Regulation Act'. This privilege is the examination by committee of themselves of the mine and its condition as regards safety."[79]

It was not simply a question of the men availing themselves of a committee, however. Dick claimed that he had "reason to believe that the managers of both of the collieries just named [Wellington and Union] would be pleased if their workmen would take advantage of the privilege referred to, and examine into the condition of the mine at least once a month,"[80] but the truth was that pit committees were never welcomed or accepted by the management of the Dunsmuir mines. Although committees were permitted by law, workers had little incentive to file grievances, especially if they lived in controlled housing; those who did were branded as troublemakers and were often summarily dismissed. Whenever the men did avail themselves of pit committees at Wellington, Cumberland, and

Extension, they were harassed. When one committee reported gas at Extension during the summer of 1912, its members were dismissed, touching off the Great Strike. Unions were also directly beneficial to underground safety because they offered protection against the notorious work/safety trade-off. "Only through the collective power of the union," writes Keith Dix, "could mine workers look forward to income levels high enough to give them the freedom to work safely."[81] Good wages, in other words, reduced the pressure to ignore dangerous work conditions and to take risks.

Finally, because of the policies of the VCMLC, the workforce at Nanaimo was more stable. This reflected not only the divergent nature of the two industrial communities but also Robins's tendency to hire more experienced workers. Due to the Dunsmuirs' intransigent policy towards unionization, they tended to hire more inexperienced, and hence more accident-prone, workers. It is difficult to say to what extent the higher accident and fatality rates at the Union mines were due to the large number of inexperienced Chinese workers employed underground (after 1889 they were no longer employed underground in Wellington/Extension or Nanaimo). European miners certainly felt that the Chinese posed a danger underground; safety figured prominently in their anti-Chinese discourse and agitation, which in turn was an important catalyst for unionization. They alleged that because of their lack of English, their "unassimilability," and their alien ways, the Chinese created insurmountable safety hazards underground.

Was the issue of Chinese safety an excuse to exclude them from the mines, as Patricia Roy has suggested – after all, white miners did not object to Belgian or Finnish miners who could not speak English – or was there an element of truth in the accusation?[82] Lambertson has stated that "whether or not Asian workers were 'really' sources of danger underground is irrelevant in understanding attempts to exclude them from the mines."[83] But clearly the point is that, whether accurately or not, European miners and society at large perceived the Chinese to be dangerous. As Belshaw correctly concludes, "concerns for workplace safety seemed measurable and demonstrable."[84]

Following the agreement to exclude the Chinese from underground work at Wellington and Nanaimo in 1888, the Chinese were employed underground only at Union, the most dangerous mine on the Island. But evidence, albeit incomplete, suggests that at times the Chinese did suffer disproportionately high mortality rates. Of the 28 deaths in the Island mines in 1903, for instance, 22 were Chinese (of a workforce of 799, for a rate of 27.5 per thousand). The following year, the number improved dramatically. Out of 9 fatalities, 4 victims were Chinese, for a rate of 5.7 per thousand, compared with a rate of 2.2 per thousand for European miners.[85] Of course, these figures do not indicate that the Chinese were inherently unsafe. On the contrary, the employment of Chinese labour was not in and of itself a reason why accident rates were high or increased; rather, as one study of American mine safety has concluded, the employment of inexperienced workers, most often hired during periods of rapid growth in production or labour unrest, tended to cause accident rates to rise. "New immigrants, who lacked mining experience and often the ability to speak English," had higher accident rates, writes Fishback. Significantly, however, these rates declined as they gained experience.

Indeed, there is little evidence to support the conclusion that the Chinese generally were inherently unsafe. Dunsmuir and his managers, John Bryden and Francis Little, certainly regarded the safety issue as an excuse to get rid of their Chinese workers. They vehemently disagreed with Robins and the European miners who suggested that the Chinese were dangerous. No. 2 mine at Union, Little argued in 1902, had been worked for eight years (illegally) by about 150 Chinese employees under the supervision of three white men, and there was "never a man killed in it." The Chinese did everything except supervise and inspect for gas, which was the job of the white overman and firemen. "We found it quite satisfactory in every way," he concluded. "I do not consider the Chinese any more dangerous than whites. I think they are a little bit more careful. They won't take risks. In case of accidents they are not a bit more subject to panic."[86]

This was confirmed by none other than Archibald Dick, inspector of mines, who reportedly claimed that "he would prefer to work with an ignorant Chinaman than an ignorant white man."[87] The exclusion of the Chinese underground at the VCMLC did not make the Nanaimo mines safer. The presence of the Miners' and Mine Labourers' Protective Association (MMLPA) and of a management whose labour relations strategy was based on negotiation rather than conflict did, however, help reduce accident rates. Generally speaking, skilled, experienced workers were less accident-prone than the unskilled and inexperienced, and organized workers were better able to break the vicious cycle that characterized the division of safety labour and the doctrine of individual responsibility. Of course, this was no guarantee of survival.

The geological peculiarities of the coalfield, the system of mining operations, the nature of class relations, and the role played by the state all determined the general level of safety in a particular mine. The conflicts and tensions created by these interconnected structural factors were revealed during the inquest into the Extension explosion of 1909. The explosion was triggered by a large fall of coal along the 2½ level of No. 2 mine, which released gas that was ignited by an open-flame lamp. The question that preoccupied the three different investigators was not the cause of the ignition of the firedamp but the source of the cave-in. Francis H. Shepherd, chief inspector of mines, and W.F. Robertson, the provincial mineralogist, maintained initially that the cave-in was the result of an undetected roll in the strata and had occurred by natural causes under normal conditions, as there was no "evidence of any sudden jar or shock in the mine at the time." Robertson even speculated that it might have been caused by an earthquake, since the seismograph at the Meteorological Observatory in Victoria had recorded "an abnormal change, due to an unusual wave in the earth's surface," on the morning of the catastrophe.[88]

The inquest agreed with the findings of Shepherd and Robertson. The massive fall of coal was the result of an undetected abnormality in the geology of the seam that had made the roof unstable. Since there was no apparent prior evidence of force, and since under these conditions no force or concussion was necessary for the roof to collapse, no blame was attributed either to the men or to the company. The final verdict of the jury inquest was therefore satisfactory to both. Shepherd and Robertson argued that the proper precautions had been taken. As they pointed out, the fireman had indicated in his record book that conditions in the mine on the morning of the explosion were good. Ventilation was adequate, coal dust had never been a serious problem because the mine was so humid, and "no explosive gas had been found for fifty-five days previous to October 5th, 1909."[89] In addition, Robertson reported that 2½ level had been timbered "in the usual way, as had been sufficient elsewhere in the mine." In fact, so soft was the coal in the roof that many timbers and stringers were left standing after the cave-in.[90] Their reports also avoided any criticism of the miners. In his annual reports, the provincial mine inspector usually blamed accidents on worker carelessness, but the investigations into the Extension explosion defended practical miners. After all, "no practical miner or shot-lighter would fire one shot while he had another charged 6 feet distant."[91]

The *Nanaimo Free Press* began speculating about the cause of the explosion shortly after the disaster. The newspaper concluded that it had been caused by a blown-out shot. While this had not yet been confirmed, the paper reported that this was Shepherd's opinion, and that "among the men this is now a well developed theory"; after all, "all miners are not good miners."[92] In the end, this explanation was supported not by Shepherd but by James Ashworth, a British mining engineer and general manager of the Crow's Nest Pass Coal Company. According to Ashworth, an improperly fired shot or blown shot in room 29 just prior to the explosion had produced the force that brought down the roof. He maintained that the body of miner Thomas Thomas had suffered burns on the hands, indicating the presence of flame, and that he had "drilled a series of three holes, and then removed his drilling tackle into the level; he then charged two holes, in contravention of Rule 9A of the 'Coal Mines Regulation Act,' and fired one, probably intending to fire the other afterwards."[93] Shepherd and Robertson, backed by the evidence of miners who participated in the investigation, refuted Ashworth's claim. Thomas, they argued, had indeed fired a shot that morning, but it had not been fired "immediately preceding the explosion." Moreover, it had been a good shot and by the time of the explosion, Thomas had already loaded almost three cars of coal and had enough unloaded coal at the face for three more cars.[94]

Although the inquest agreed with Shepherd and Robertson, its conclusions were not very satisfying. The doubts raised by Ashworth's theory lingered in the minds of many. The problem was not finally solved until June 1910, when Manuel Delcourt, driving a new stall next to James Keserich's former place, holed into the face of room 27 and discovered evidence of a "badly planted and partially blown-out shot." This new evidence, which was immediately reported to the inspector, was the "missing link." A blown-out shot had indeed caused the explosion, but it had not been fired by Thomas in room 29, as Ashworth had thought, but by Keserich in room 27.[95] This explained the initial concussion that brought

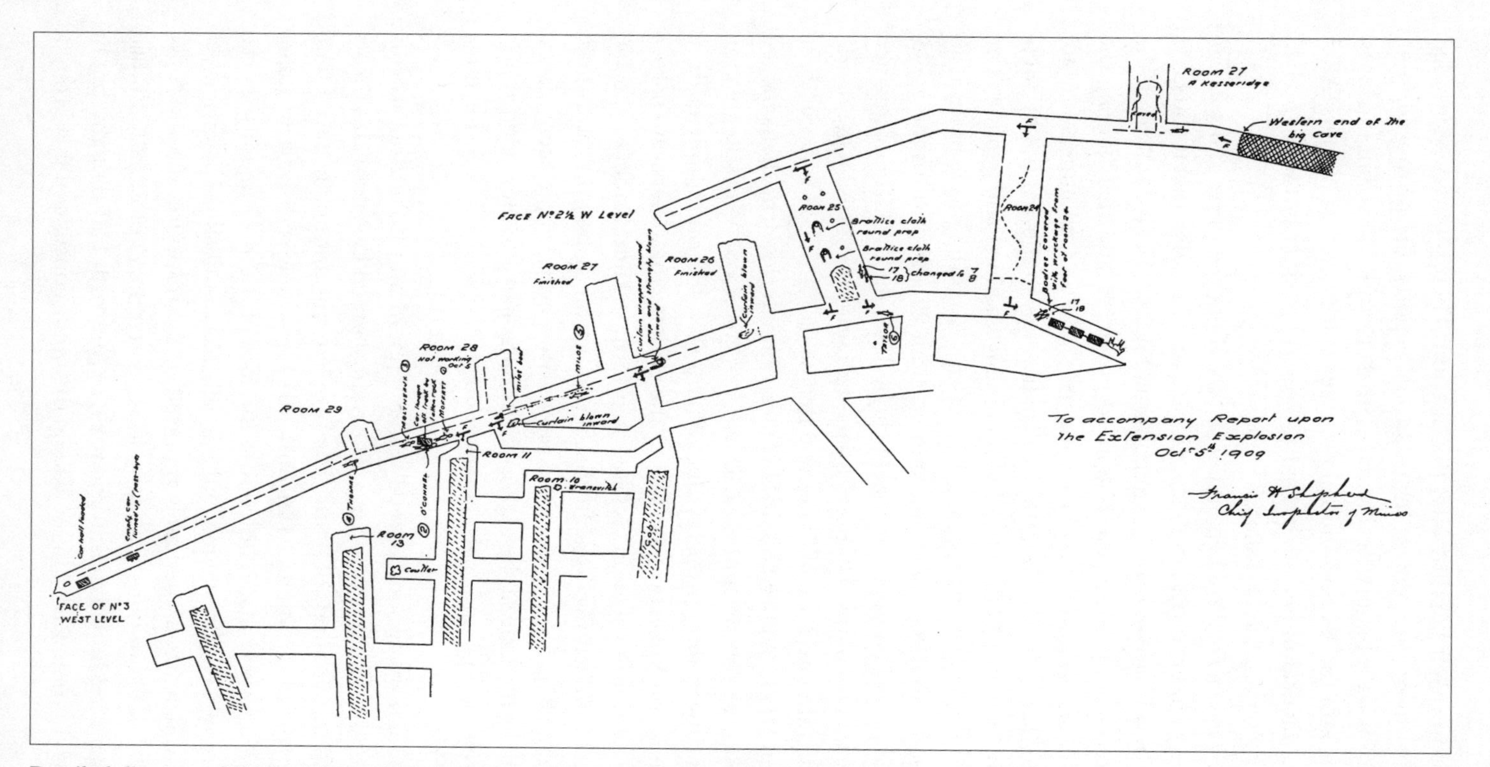


Detailed diagram of 2½ level from the investigation into the explosion of 1909. The diagram indicates the rooms and the location of the bodies. The faulty shot was eventually discovered to have been fired by James Keserich in room 27 (top right). This diagram also provides a good plan for the pillar and stall method of extracting coal. From *Annual Report of the Minister of Mines,* 1910.

down the first cave-in, and why the force of the explosion (or explosions) traversed the two districts of 2½ level.

Although the official report exonerated the miners, Shepherd's new report suggested that miner carelessness had caused the explosion. This, it seems, had not been far from the investigators' thoughts during the initial inquiry. Despite the fact that Robertson believed that "shot-firing had apparently nothing to do, in any way, with the explosion," in his final recommendations he claimed that the miners did engage in highly dangerous practices while shotfiring and accused them of refusing to follow standard mining regulations. He had found evidence, for example, that the shotlighter did not always examine the hole before it was charged "to see that the coal was well prepared, the hole properly placed, cleaned, etc.," and that there was "some doubt whether the miner having a hole ready and charged always waited for the shotlighter to come around, but fired his own shots when he got ready, against orders and regulations."[96] John McMurtrie, a fireman at Extension, revealed that the men did not wait for a shotlighter because no shotlighter had been appointed; the men did their own firing. This statement was contested by Andrew Bryden, who claimed that they were obliged to have a shotlighter and that he enforced the law.[97]

In fact, Shepherd's supplementary report was somewhat ironic. He had to explain why the investigators and miners had not found this evidence earlier. This must have raised new questions about the quality of the investigation and the inspections, which did not seem to be that high. For instance, on 6 September 1909, one month before the explosion, Archibald Dick, the inspector of mines, had visited Extension. He reported that he had found no trace of gas and that ventilation underground was good, although he could not test the quantity of air because, he wrote, "I had omitted to take my instrument with me."[98] Whether or not this was a contributing factor, it was nonetheless a revealing admission. And once again, no mention was made of the company's responsibility for supervision and enforcement of existing regulations.

The inquest into the Extension explosion offered little comfort to the miners and their families. The investigators could not accurately pinpoint the actual cause, and their conflicting interpretations undoubtedly added to the confusion and uncertainty that surrounded explosions. As the

*Nanaimo Free Press* remarked shortly after the Extension explosion, anticipating in a way the outcome of the inquest: "In many cases the cause of these mine catastrophes remains a perfect mystery, and the holocaust of miners goes unexplained."[99]

Mystery or not, no miner was indifferent to the dangers he faced on a daily basis. Safety was a major concern, but clearly cannot be understood in isolation from economic and political factors. Miners' struggles for improved working conditions, better wages, and union recognition cannot simply be reduced to questions of workplace authority and autonomy. Their much-touted fatalism notwithstanding, the miners' fight for legislation, regulation, and unionization reflected their conviction that reforms would improve not only living standards but also their physical health and welfare. Safety was consequently an important factor in their political and economic mobilization. In the wake of the 1887 and 1888 disasters in Nanaimo and Wellington, the miners intensified their demand for better regulations and reform. Although they maliciously and falsely blamed the Chinese for the disasters, safety figured prominently in their anti-Chinese discourse and rallied public opinion to their cause. This period of mobilization would culminate in the formation of the Miners' and Mine Labourers' Protective Association, the first successful union in the Island coalfields. It is no coincidence that in the wake of the Extension explosion, the United Mine Workers of America (UMWA) was invited to organize the Island miners.

But the impact of death and injury was felt in other ways. On the one hand, it strengthened the bonds of community solidarity. Thus, when Frederick Greaves died from injuries sustained by a fall of top coal in September 1908, the entire community mourned his tragic loss. A miner active on the local school board and in civic affairs, Greaves was eulogized as a man "held in high esteem by all classes." On the day of his funeral, businesses closed, and the St. John's Lodge and the Masons, which provided the pallbearers, led highly ritualized processions through the town accompanied by the town band, as befitted the heroic celebration of Greaves's death.[100] At no time was this more true than in the wake of the 1909 explosion. While death and injury were common occurrences in the mines, nothing can prepare people for the trauma and consequences of such a catastrophe. Few residents of the town were untouched by the calamity. Thirteen women were widowed, and thirty-eight children, only two of whom were of an age to provide for themselves, lost their fathers.[101] The community, plunged into mourning, mobilized to assist the families of the victims. Following a public meeting, the town council established a relief committee, headed by Mayor Daniel Nicholson and the bank manager, L.M. de Gex, which organized and distributed the relief funds that poured in from various individuals, groups, and municipalities throughout the province.[102]

Just as the explosion killed indiscriminately, regardless of ethnicity or religion, so too community relief, although manifestly inadequate, did not discriminate. Ethnic, religious, and class barriers were broken down as the town buried its dead; businesses remained closed on the Thursday and Friday for the funerals that were attended by the general public. The various fraternal orders and friendly societies were prominent in the relief effort. When the bodies were brought from Extension to Ladysmith, "some of the representatives of the secret orders were present [at the train station] to see that the members of their order were accorded the attention usually given to a dead member of their society."[103] Members of the Hospital Auxiliary as well as other groups also assisted the families of the victims in many ways, offering comfort, minding children, and looking after their immediate material needs.

On the other hand, death and injury reinforced the occupational community. Miners were at or near the bottom of the prevailing class structure. Wages, while comparatively high, did not compensate for this, and society at large had little regard or sympathy for coal miners' aspirations and struggles. While most people might have winced at the conditions in which miners were obliged to work, the more conservative middle-class public, as numerous accounts testify, viewed miners with a mix of distaste and fear. The very disturbing and menial nature of the job – working semi-naked in the dirt and darkness miles underground – seemed to set miners apart from the rest of society. Uncouth, ignorant, brutalized, drunken, violent, and menacing, miners were often regarded as "a class of people who were little removed from barbarism, and whose home was down in the eternal darkness."[104]

Given prevailing attitudes and the harsh nature of the work, miners developed strong bonds that resembled a closed fraternity. While they knew that mining was not the easiest form of employment and that it took a heavy toll on bodies and lives, many miners viewed it positively. They often came from mining families and had been "socialized as boys in the pit."[105] For men such as Bill Johnstone, a miner who emigrated from the north of England to Canada in the 1920s, coal dust was "in his blood." Accordingly, "by tradition [a man] was a coal miner. His thinking was conditioned to coal mining and that was the only skill he thought he had."[106] His level of skill, knowledge, and craftsmanship, upon which not only production but his very life depended, was a source of pride and his work sometimes a source of pleasure.[107]

One of the most compelling reasons for the high degree of occupational loyalty was that the coal-mining fraternity fostered deep bonds of friendship that found expression in a common identity or outlook. This occupational community was born of the shared experience of work, danger, and death; the miners' prevailing sense of pride and tradition; and a deep, historical awareness of their unique position in modern industrial society, all of which went largely unnoticed by society as a whole. This identity was also forged through struggle against management and the state, and against the pressures that threatened a culture and a way of life. The bonds of solidarity and independence empowered the miners during a time when their industry was in transition and when their status in the workplace and society was under siege.

# 5
# *From Pillar to Post*

In his testimony to the 1903 Royal Commission on Industrial Disputes in the Province of British Columbia, James Dunsmuir declared that he had made his workers move to Ladysmith so that they would be further removed from the unionized mines in Nanaimo.[1] His hopes of avoiding labour unrest were soon dashed, when on 12 March 1903 Ladysmith miners on the afternoon shift refused to board the train to Extension, beginning the Island's first major colliery strike of the new century. The 1903 strike was a watershed in the history of working-class mobilization on the Island, as the miners renewed their efforts to organize and to win union recognition, this time under the banner of an international industrial union, the Western Federation of Miners (WFM). The miners did not emerge victorious, but the strike represented the beginning of a new style of labour agitation that shook the province's ruling classes to the core. Frightened by the role played by the WFM and by the increasingly articulate political voice of the masses, symbolized by the growing strength of socialism and social democracy – the first socialist MLAs were also elected in 1903 – both business and the state responded with an aggressive campaign against organized labour and the working classes, an offensive that reached a climax on the Island during the protracted and violent strike of 1912-14.

By 1903, according to the conventional historical narrative, the labour movement had completed a fundamental shift away from moderate political reformism or labourism towards a more extreme or "impossiblist" style of Marxist socialism inspired by the "frontier labourer." Whereas skilled, organized urban craft workers tended to act as a conservative force, unskilled resource workers, poorly organized, increasingly the subject of state repression, and victimized by new techniques of managerial control, experimented with new forms of organization and embarked upon aggressive job actions. The coal miners of Vancouver Island, considered by some to be inherently "radical" – after all they were the archetypical "frontier labourer" – were thought to be instrumental in the radicalization of the labour force.[2]

There is little doubt that the pace of working-class political and economic mobilization in the coal-mining communities of Vancouver Island accelerated after the turn of the century, but the standard image of miner radicalism and militancy has to be revised if not discarded altogether. The electoral success of the socialists, for instance, or the popularity of the WFM and later the United Mine Workers of America (UMWA), is by no means proof that reformism had died or that the miners, who viewed themselves as independent, skilled craftsmen – not unskilled labour – had abandoned their essentially conservative principles in favour of revolutionary Marxist socialism. The rhetoric of class and class conflict did serve as an important integrating force within the community, but the political landscape remained diverse, the left wing fragmented and weak, and, if we go beyond the rhetoric, the miners still looked to the legislature and the trade unions for concrete reform. Reluctant to strike, the coal miners were almost always prepared to seek accommodation with capital. Despite the electoral success of socialism in the coalfields, continuity, rather than fundamental change, was the main characteristic of working-class politics on the Island. Although the miners were greatly concerned about their position and power in the existing capitalist order – as Seager has pointed out, "workers did not willingly acquiesce in a system in which they were the losers"[3] – their objective was less to overthrow that order than to redefine their position within it.

During the first decade of the twentieth century, there was a marked increase in the pace of unionization and the number of strikes in the industrialized world. In Germany, for example, the number of strikes more than doubled, from 1,468 in 1900 to 3,228 in 1910. France averaged 500 strikes per year between 1900 and 1915, and in Britain the annual number of work stoppages also doubled between 1907 and 1913. This period also saw an intensification of labour activism and conflict in Canada, as strikes "became something of the norm in the social relations of production."[4]

Historians have identified three major strike waves during the first two decades of the twentieth century: 1899-1903, 1910-13, and 1917-19. In his important study of strike activity in Vancouver, James Conley has shown that these years "accounted for 75 per cent of the strikes, 94 per cent of the strikers, and 91 per cent of the striker-days in the Vancouver area,"[5] a pattern similar to that of the Maritimes and Ontario. At the forefront of this strike activity were skilled craft workers in the manufacturing or construction sectors, but coal miners also figured prominently in these periods of intensified class conflict. It would be a mistake, however, to argue that coal miners were especially "strike-happy" or more radical than other workers.[6] On the contrary, they engaged in fewer industrial disputes than other trades, but because their strikes involved large numbers of men and often lasted longer, and given the general importance of coal to the modern industrial economy, they usually had a greater impact on society and were of greater concern to authorities.[7]

Despite regional variations and industrial peculiarities, most of the strike activity also took place during times of economic growth and prosperity. When the market for coal was strong and profitable, when demand for labour was high, and when companies could not risk finding themselves without a secure workforce, miners often won important concessions from the companies. In Britain the strike waves of 1889-93 and 1910-13 corresponded with a tight labour market, and in Germany strikes followed "in the wake of periods of a long and sustained rise in working-class living standards, as in 1889 and 1912."[8] The same was true in Canada, although there were of course notable exceptions to this rule. McKay points out that although "strikes were most common in times of prosperity," Nova Scotia miners waged their largest, most impressive strike during the "depths of recession" from 1909 to 1911. Not surprisingly, "they fell almost silent after their defeat." The Vancouver Island strike of 1912-14 began just before recession hit the region, and in the wake of the defeat, the Island miners also fell silent.[9]

Disputes between labour and capital accompanied the development of the coal-mining industry on Vancouver Island, although this had less to do with the mythical radicalism and militancy of miners than with appalling work conditions and their understandable determination to protect their economic interests and social rights against the inherently exploitative forces of predatory capitalism. John Belshaw has calculated that there were at least twelve "noteworthy strikes and lockouts" in the Vancouver Island coalfields between 1850 and 1914.[10] Owing to the small population and the primitive nature of the early mines, labour unrest during the 1850s and 1860s was sporadic, characterized by short work stoppages and ineffectual protests. The first recorded incident occurred in 1850, shortly after miners arrived at Fort Rupert from Britain. Led by the irascible Scottish miner Andrew Muir, the men struck over intolerable work conditions and wages. They were isolated and few in number, and the strike ended soon enough when the Hudson's Bay Company (HBC) placed Muir and fellow miner John McGregor in irons.[11] The work stoppage, like the attempt to dig coal at Fort Rupert, ended in dismal failure.

This was a minor dispute, but it nonetheless established the pattern of class conflict that would last until well into the twentieth century. On the one hand, the miners were willing to dig coal and reluctant to put down their tools, but they demonstrated that they would fight to improve working conditions and wages and would defend their perceived rights and dignity as skilled labourers. On the other hand, the Hudson's Bay Company revealed that it would not tolerate any challenge by the workers to its authority and to the status quo, and that it would use force and violence to control them, a strategy of labour relations that would continue under the autocratic and paternalistic regime of the Dunsmuirs and the large corporations.

In the two decades following the closure of the HBC's small works at Fort Rupert and the beginning of operations at the extensive Nanaimo coalfield, both production and the size of the workforce steadily increased. Labour unrest also became more common. The first major strike occurred in 1864 over wages at the Vancouver Coal Mining and Land Company (VCMLC), which had bought out the HBC two years previously. Because of the general lack of skilled labour, the miners felt that they were in a relatively strong position during the lengthy dispute, which lasted from August 1864 to February 1865. The position of the striking miners was seriously undermined, however, when for the first time the company recruited Chinese strikebreakers, many of whom had recently arrived from the gold fields of the mainland. This strike was followed by a five-month stoppage at the VCMLC in 1870-71 over wages.[12] Despite the willingness of the miners to stop work, no effective form of organization emerged. This was not due to the lack of interest or effort on the miners' part, although clearly not all miners supported unions, but rather to the operators' intractable hostility towards all forms of miner organization and because the workforce still remained quite small and fragmented.[13]

The first union on the Island, the Coalminers' Mutual Protection Society, was formed during the 1877 strike. Although it attracted many of the white miners from Wellington, Nanaimo, and the smaller mines in Harewood and Chase River, it could not withstand the combined onslaught of the companies and the state and did not become a significant force. Politically and economically peripheral, it existed only clandestinely until April 1883, when it re-emerged publicly as the Miners' Mutual Protection Association (MMPA).[14] The MMPA has been correctly called an "embryonic trade union ... an organization that bridged the gap between benevolent association and trade union."[15] However, it folded during the 1883 strike at Wellington. As had become the pattern in previous disputes, Robert Dunsmuir fired union men, hired Chinese strikebreakers, and used his political connections in Victoria to requisition the militia and special constables to intimidate and coerce his employees. Nonetheless, the MMPA laid the groundwork for future organization and, more importantly, began to redefine the terms of industrial conflict on the Island. When the first local of the Knights of Labor was established in Nanaimo in December 1883, its organizers found a workforce receptive to its emancipatory program.

The arrival of the Noble and Holy Order of the Knights of Labor on Vancouver Island marked an important turning point in the mobilization of the coal-mining population. Founded in Philadelphia in 1869, the Knights of Labor embraced all workers, rather than just the skilled tradesman. If the Knights' "effectiveness stemmed from its ability to build on the mundane class distinctions of daily life and to construct out of this a movement that attempted to unite all workers to oppose the oppression and exploitation that they lived through both on and off the job,"[16] its attractiveness was in large part due to its conscious attempt to revitalize the values of an idealized and romanticized lost age. As its name suggests, and as its ritual and symbolism seem to confirm, the Knights resembled a premodern, corporatist order, unsure how to confront the exigencies of modern industrial capitalism. It would be a mistake, however, to infer from this apparent contradiction that the movement was backward-looking or anachronistic and hence doomed to failure. On the contrary, by constructing a public image of the worker as a noble, dignified, and autonomous member of society with common interests and ambitions, as distinct from the world of the middle class, the Knights contributed greatly to the formation of working-class consciousness. Of particular importance in the coalfields was the fact that the Knights recruited skilled colliers and unskilled labourers, creating an unprecedented level of unity and solidarity among the very diverse elements of the mining population. According to historian Bryan Palmer, the Knights helped refashion a largely fragmented and inchoate working-class experience into "a movement culture of opposition and alternative, a process of working-class self-activity that took the collectivist impulses of labouring experience and shaped them into a reform mobilization."[17] The importance of this should not be minimized. As Belshaw has correctly argued, "the Knights of Labor paved the way for the industrial unionism and political activism of the early 20th century."[18]

The Knights enjoyed considerable success in British Columbia in the 1880s. Local assemblies existed in all of the major centres of the province. By mid-1884, barely half a year after its arrival in the Island coalfields, there were 241 members in the Nanaimo local. Within a short span, the organization had spread to Wellington and North Wellington, supplanting the MMPA.[19] The Knights' success was relatively short-lived, however, and by the end of the decade support began to decline. This was the result of a number of factors. The emergence of working-class political parties, a reflection of the divergence between the interests of labour and society at large; a new sense of worker autonomy; new pressures from a state determined to control labour activism in order to foster economic expansion; and the increased popularity of trade unions promoting a more restricted and disciplined form of organization all undermined the viability of the Knights. Broader, long-term changes in the structure of capitalism – the intensification of the process of monopolization, rationalization, and consolidation – also contributed to the decline of the Knights, because they inevitably increased the polarization of labour and management and led to new sources of class conflict.[20] Within this changing context, a new organization was eventually formed in the Nanaimo coalfields in February 1890, the Miners' and Mine Labourers' Protective Association (MMLPA), which soon superseded, but did not replace altogether, the Knights of Labor.[21] More concerned with practical political and economic issues, such as union recognition and a living wage for its members, than with articulating an alternative working-class culture, the MMLPA succeeded in recruiting members in Nanaimo and Wellington, with more than one thousand men, almost the entire European workforce, joining by May 1890.[22]

The MMLPA was unusual in the history of unions in the Island coalfields because it was the only union to win official recognition. Its victory was far from complete, however. Samuel Robins, manager of the VCMLC in Nanaimo, agreed to recognize the MMLPA in May 1890, but James Dunsmuir never overcame his ideological intransigence and adamantly refused. While Robins eventually signed an unprecedented closed-shop agreement with the union the following summer, Dunsmuir fought successfully against his miners' demands for union recognition during the bitter twenty-month strike at Wellington in 1890-91, dismissing those involved in union activity.[23] Ironically, the MMLPA's success in Nanaimo was possible only because of the stand taken by Robins. In the wake of the Wellington strike, the MMLPA was on the verge of collapse, the number of lodges in the district having declined from four to two. During this low point in the union's fortunes, Robins recognized it as a bargaining agent.[24] Robins has been described by some as "half-progressivist, half-traditional," and by others as paternalistic.[25] Whichever was true, his belief that "the true interests of capital and labor are absolutely and unalterably identical"[26] was ultimately a reflection of his pragmatism. His efforts to revitalize and sustain the MMLPA were designed first and foremost to improve the position of the company. Negotiation and cooperation, rather than confrontation, would reduce friction, improve the stability of the workforce, and hopefully increase production and profits.

In the long run, Robins's instincts proved correct. The MMLPA was a company union, later condemned by some as a "figurehead,"[27] whose survival depended on its "being vital to the VCMLC."[28] Radical and militant impulses found no home in the MMLPA, and a very close relationship developed between management and union leadership, to the obvious benefit of the company. Not only was Robins able to keep wages down – for instance, in 1893 he negotiated a 20 percent wage reduction – but he was also able to avoid serious work stoppages throughout the life of the union.[29]

Historian Jeremy Mouat has suggested that the strike of 1890 and the subsequent success of labour politics under the auspices of the MMLPA marked a turning point in labour mobilization. In particular, he argues that during the 1890s the miners became radicalized and that "working within the Legislature was discredited."[30] Despite the success of the MMLPA at Nanaimo and the election of a number of labour candidates, the most dominant feature of working-class and labour history during the decade of the 1890s was moderation rather than militancy. The miners' union, like the Knights of Labor before it, placed a premium upon labour peace, avoiding potentially harmful disputes and opting to strike only as a last resort. There is also little evidence to suggest that miners, unionists, or labour politicians viewed the legislature as discredited. The labour movement measured its success or failure in terms of the number of members it returned to Victoria and in their subsequent ability to pass legislation beneficial to the working population. For many years, the miners had looked to Victoria in an effort to improve working conditions and mine safety. Knowing that company owners resisted improvements and balked at legislation, they protested, signed petitions, and eventually elected MLAs to fight their cause. And it was precisely in the legislative realm, rather than in the extra-parliamentary sphere of union activism, that the miners registered their most significant victories, as can be seen in the host of laws passed over the decades, the most important of which included the Coal Mines Regulation Act of 1877, the exclusion of the Chinese underground in the late 1880s, the eight-hour law, and eventually the Workmen's Compensation Act of 1902.

The MMLPA became a significant and influential force in the lives of its members and in local society, but its moderate reformist agenda, rather than reflecting the radicalization of the mining population, revealed the limits of working-class mobilization and militancy in the 1880s and 1890s. Miners did confront the company and oppose state policies, but many also clearly felt that they had a legitimate stake in the existing economic order. This was especially true of the skilled colliers or the "labour aristocracy," whose status as craftsmen served as a bridge between the working and middle classes and who often engaged in "penny capitalism."[31] Indeed, labour politics in the 1890s were not characterized by an anticapitalist rhetoric but by specific, locally derived populist issues that appealed to a variety of political ideologies. Chinese exclusion and the critique of large landowners and monopolists – particularly the Dunsmuirs, who limited not just competition in the coal industry but also economic opportunities generally through their control of land and resources – were cross-class issues that dominated political discourse in the coalfields and generated considerable support from the small-business community.

What this meant in terms of political development has been demonstrated by John Belshaw, who has provided the most astute analysis of working-class political activity in the coalfields during the period between 1870 and 1900. The existence of an influential elite of skilled miners, who viewed themselves as craftsmen and contractors of labour, combined with real possibilities of social mobility, reinforced the miners' conservatism. There is perhaps no better indication of the political will of the miners than election results, and as Belshaw points out, the polls from these years demonstrate not the inherent radicalism of the miners but rather clear "evidence of working-class conservatism."[32] Despite the fact that both the Knights of Labor and the MMLPA nominated candidates for the provincial legislature under a variety of labour banners, it was by no means automatic that the miners would support working-class candidates. In 1875 and again in 1894 and 1898, John Bryden, Dunsmuir's manager, was elected in Nanaimo, as was Robert Dunsmuir in 1882 and 1886, and James Dunsmuir in 1900. These three staunchly conservative, anti-labour men succeeded in gaining the trust of the miners – although intimidation in a few individual instances cannot be entirely ruled out – whereas labour candidates continuously faced an uphill battle.

The first working-class representatives, Thomas Keith, Thomas W. Forster, and Colin McKenzie, were not elected until 1890.[33] They did not have established political parties, and their platform, based upon that developed by the Workingman's Party for the 1886 election, reflected divergent interests and was hardly the stuff of militant socialists. Rather, their program was designed to have a broader appeal, combining criticism of large land grants and monopolies with specific demands for safety legislation in the mines, Chinese exclusion, and the eight-hour workday.[34] Their success was short-lived. In the election of 1894, the three Nanaimo-area candidates, Thomas Keith, Tully Boyce, and Ralph Smith, all of whom were nominees of the MMLPA-dominated Nanaimo Reform Club, failed to win their seats.[35] This reflected the weak position of the MMLPA and the fact that the miners did not define their interests simply according to class. Smith lost by a huge margin to Dunsmuir's manager, John Bryden, who garnered almost 75 percent of the vote in his North Nanaimo riding. Only after the turn of the century would working-class political mobilization produce consistent and positive results at the polls.

The limited success of labour candidates in provincial elections prior to 1900 exposed significant divisions within the working class. The growing mining population was not a cohesive, homogeneous group. The miners were divided according to skill level and ethnicity. While the immigrant population grew and the percentage of British-born workers shrank in relative terms, new Europeans were gradually integrated over time.[36] The same was not true, however, of the growing Asian population.

Although Chinese exclusionism was only one plank in the working-class platform, it was perhaps the dominant political issue in the 1880s and 1890s. In the absence of any clearly articulated and overriding ideological principles or class sentiment, anti-Chinese agitation was of fundamental importance. It was the one issue that united the socially and ethnically diverse population of European miners and mine labourers. Equally significant, it created important linkages between the working classes and their middle-class counterparts in Victoria and Vancouver, lending their exclusionism an aura of respectability.

But if exclusionism was the glue that initially bonded the miners together, it held little promise for the long-term development of working-class economic and political power. Class conflict was not just "confused by the issue of race and the consciously constructed notions of 'white work' versus 'oriental labour,'" as Palmer has suggested; it was seriously undermined and subverted. Although the Knights of Labor, for instance, pursued an active agenda, including the struggle for shorter work hours, political reform, and the end of monopolies, its success largely hinged upon its willingness to exploit the existing racial tensions between European and Chinese miners. Talk of working-class unity notwithstanding, the Knights' corporatist vision of a community of labour based on the shared experience of oppression and exploitation could never have been realized because it rested on racial exclusion. The racial issue was not introduced by clever capitalists following a conscious policy to divide and conquer the working class by diverting class conflict into racial conflict. On the contrary, racism was an essential component of the European worldview and a vital part of "popular white working-class mythology."[37] Refusal to include the Chinese, which encouraged the employment of them as strikebreakers, was the Achilles' heel of the labour movement in BC. It destroyed any hope of working-class solidarity and prevented workers from addressing the real causes of their oppression and exploitation.

Although the 1890s began with a major strike, the decade was one of relative labour peace in the Island coalfields. The embryonic trade unions and mutual societies like the MMLPA remained weak. Labour politicians lobbied for reforms, but the political voice of the working class was fragmented, poorly articulated, and easily silenced by conservative political and business interests. This changed during the first few years of the new century. Throughout the province there was a surge in working-class organization and activity, culminating in the strike wave of 1903.

During this tumultuous year, labour strife affected most nations of the industrial world. It peaked in France, Belgium, and other industrialized nations of Western Europe. Commenting on strike waves in the Netherlands, Italy, and Russia, the *Nanaimo Free Press* ran a headline declaring: "Workers Waging World Wide War."[38] This was the case closer to home as well. Unionized (WFM) miners in the East Kootenay communities of Coal Creek, Michel, and Fernie walked off the job on 11 February, followed a few days later by a short-lived strike at the Nanaimo collieries. The most important strike, however, began on 27 February, when clerks, freight handlers, and labourers in Vancouver, members since 1902 of the United Brotherhood of Railway Employees (UBRE), took on the mighty Canadian Pacific Railway (CPR) in response to the dismissal of a union employee.[39] The CPR's intransigence and determination to break the union ensured that the strike quickly spread down the line from Vancouver to Revelstoke, Nelson, Calgary, and Winnipeg. In the end, the corporate power of the CPR proved too much for the unions. As McDonald concludes: "Faced with aggressive CPR tactics, including espionage, the importation of strike-breakers, and the murder of labour leader Frank Rogers, the Vancouver-centred strike collapsed, destroying the UBRE and three other unions with it."[40]

Shortly after the railway workers walked off the job, miners at Extension went on strike, leading to the accusation that they were involved in a sympathy strike. Newly affiliated with the WFM, which, like the UBRE, was a member of the politically engaged American Labor Union (ALU), the Extension miners knew that the strike would disrupt the delivery of coal to the CPR. Fearful of sympathy strikes, an unambiguous statement of union and working-class power, the authorities subsequently led a frontal assault on the strikers and their unions. As is clear in the detailed report of the Royal Commission of 1903, they viewed the Extension strike as part of a wider conspiracy by illegitimate revolutionary socialist organizations affiliated with American labour interests to disrupt the Canadian economy and to corrupt Canadian workers, and were therefore determined to crush the union. This extraordinary conclusion was as unfortunate as it was predetermined by the commission, and bore little relation to the evidence presented by the witnesses. It unjustly vilified the WFM and completely ignored the legitimate and long-standing local grievances of the miners over wages and work conditions, which had led to organization and to the strike mandate in the first place, and did not mention the pressing concern of mine safety.[41]

The resolution by the Ladysmith miners to join the WFM was not a coincidence or impulsive decision. Nor was it the work of "foreign socialistic agitators of the most bigoted and ignorant type," whose "primary object and common end ... is to seize the political power of the state for the purpose of confiscating all franchises and natural resources without compensation."[42] Rather, the men were motivated by local issues, the most important of which was protection against the arbitrary measures of James Dunsmuir. This was recognized by many, including the writers at the *Nanaimo Free Press,* which declared that "in compelling his men to move to Ladysmith, Mr. Dunsmuir sowed the seeds of the present crop of trouble which he is reaping. Driven from pillar to post, abandoning at his dictation homes which had cost years of self-denial to erect, it was but natural that they should form a union to safeguard themselves against further aggression." The WFM was the only viable option for the miners. "Local unions," the newspaper continued, "already had been proved useless, amalgamation with other island unions had been prevented by Mr. Dunsmuir himself, and the only thing left for the miners to do was to ally themselves, since no purely national miners' union exists in Canada, with one of the great international organizations."[43]

Moreover, the WFM had been successful in its efforts, organizing miners not just in the Kootenays, but more significantly for the Ladysmith men, the miners closer to home at Nanaimo. In the fall of 1902, the MMLPA had voted to affiliate with the WFM as Local 177. The miners regarded the WFM as the "securest and most convenient body to join," and between November 1902 and March 1903, two thousand men in the district reportedly became members of the union.[44] The first test came on 11 February 1903, when the WFM went on strike at Nanaimo over the elimination of an allowance of twenty-five cents per day that had been paid to miners using safety lamps, and over the wage scale for longwall work.[45] The strike, which lasted only ten days, was an important victory for the new local. The company not only acceded to the miners' immediate demands but, more importantly, agreed to recognize the union. Although Thomas Russell, the Western Fuel Company superintendent, did not believe that unions were of any great benefit to employers, he understood the advantages for the workforce and, like his predecessor Samuel Robins, did not strenuously object to the presence of the WFM at Nanaimo. Not only did he negotiate an end to the short dispute with a union committee but the company checked off union dues and began drawing up a collective agreement with the men.[46]

This attitude contrasted sharply with that of James Dunsmuir, who persistently opposed unions. When the Ladysmith miners, buoyed by the success of the union at Nanaimo, looked to form their own WFM local and to fight for union recognition, they encountered determined opposition. On Sunday, 8 March, between 300 and 400 miners attended a mass meeting at the Finn Hall in Ladysmith, where they unanimously decided to join the WFM. On 10 March, Dunsmuir responded by dismissing four of the most prominent miners at the meeting, including Samuel Mottishaw Sr., his son, S.K. Mottishaw Jr., James Pritchard, and Robert Bell. The same day, the Wellington Colliery Company posted a notice at the mines and the main depot declaring that the Extension mines would shut down on 1 April and that the men had to remove their tools by that date. Faced with this prospect, the miners held another mass meeting in Ladysmith on 12 March to discuss the dismissal of the four men and the pending shutdown. The miners resolved to strike, anticipating the April lockout, and refused to approach the company until they were formally organized. This happened three days later with the arrival of James A. Baker, Canadian organizer of the WFM and the man who had brought the Nanaimo strike to a successful conclusion.[47]

The Extension mines were closed from 12 March to 3 July. Despite the efforts of local union officials to reach a settlement, Dunsmuir adamantly refused to deal with the committees and was content to wait out the miners.[48] In the meantime, union officials travelled to Cumberland, where they attempted to recruit members. Despite the Royal Commission's claim that the Cumberland men had little desire or reason to join the WFM, on 5 April they formed WFM Local 156. Instead of locking out the miners, over the course of the next few weeks, John Matthews, the manager at Cumberland, attempted to get rid of the most prominent union activists by refusing to rehire them when their places ran out.[49] His sole purpose, he later testified, was "to break up the union."[50] Throughout April, the situation in Cumberland deteriorated and there was much talk of a strike. On 20 April, Ladysmith union leaders requested the Cumberland union to join the strike.[51] The union leadership in Denver supported this and also suggested that an attempt be made to organize the Chinese and Japanese miners, giving ammunition to later conspiracy theories.[52]

Initially there was little support in Cumberland for a sympathy strike, and none whatsoever for admitting Chinese and Japanese employees into the union. At a joint meeting of the union's executive in Nanaimo on 25 April, neither issue appeared to have been discussed, and Baker's telegram to Denver requesting permission to call out the Cumberland men was made public only during the investigation of the Royal Commission.[53] Contrary to the conclusions of the commission, the decision to strike at Cumberland was not the result of outside pressure but developed a momentum of its own in response to management's anti-union strategy following the formation of the WFM local. On 1 May a three-man union committee asked for the reinstatement of the dismissed men, but Matthews summarily refused this request, claiming that this would be tantamount to union recognition.[54] At a hastily convened union meeting the next day, members were asked to vote on "whether or not we stop work until such time as the officials and members who have been discriminated against by the Wellington Colliery Company are reinstated in their proper order, with full recognition of the Western Federation of Miners."[55] The motion was passed by a majority count of 208 to 12.

Although Dunsmuir threatened to shut down his Cumberland operations if the miners walked off the job, he did not do so, unlike at Extension. There were a number of reasons for this. According to the *Victoria Daily Times,* "under the present conditions of the coal market," coal operations at Extension were not profitable. As a result, "the closing of the mines under these conditions will hardly represent a loss to the owners." The Cumberland mines, in contrast, not only produced better-quality coal and the profitable coking coal but were producing enough to satisfy the existing demand.[56] That the Cumberland coal could be mined more profitably than Extension coal was due in no small part to the fact that most of the Cumberland workforce consisted of poorly paid Chinese and Japanese labourers. Not only was this the key to the colliery's profit margins but it was ultimately Dunsmuir's trump card in the labour dispute. Although most European miners, including a number of fire bosses and shotlighters, supported the strike, and the Japanese mine labourers refused to work – many were said to be "leaving for the Fraser and other points" – between 325 and 350 Chinese workers, excluded from the WFM, continued working.[57] The company actively sought to increase the number of Chinese labourers working underground, although this was illegal. In secret sessions, the provincial examining board issued mining certificates to a number of Chinese men, enabling them to dig coal.[58]

These developments seriously jeopardized the strike but did not dampen the miners' determination. Although many young men and a few families left Ladysmith in search of work, those who remained – referred to in one report as the "home guard" – were able to count on union and community assistance. Indeed, the community appears to have stood firmly behind the striking miners. "The men are being strongly backed by the moral support of almost every citizen who has had any extended acquaintance with the drift of affairs here," commented a reporter for the *Victoria Daily Times.* Support for the union cause also crossed ethnic lines. "Perhaps the most striking feature of the men's combination," continued the newspaper, "is the almost rabid firmness of the 'foreign' element. They are the most utterly uncompromising of all," a fact attributed to their forced removal from Extension.[59]

Several factors contributed to the collapse of the strike. The WFM had not reciprocated the miners' commitment to it, probably because it was negotiating the division of jurisdictions with the United Mine Workers of America, which would henceforth devote itself to organizing coal miners while the WFM concentrated on hard-rock miners. In the meantime, the WFM did not provide enough strike pay to sustain the families of the locked-out miners. When the locals ran out of funds in May and sent an urgent request for $14,000, WFM headquarters in Denver ignored the plea.[60] This was a grave turn of events for the miners. Parker Williams, then a representative of the local union, predicted that lack of financial support for the striking miners would spell the end of the union on the Island.[61] When confronted with this issue a few weeks later, a representative of the WFM suggested that the men "appeal to Vancouver and other unions."[62] People began to speculate that the strike would soon end, and Dunsmuir, taking advantage of the uncertain situation, proposed to reinstate the men without discrimination on condition that they abandon the WFM and accept a new contract. Under the new terms, the measure of a ton of coal – according to which the miners were paid – was increased from 2,240 to 2,500 pounds, in effect a wage reduction of 12 percent. "The proposals were refused without a dissenting voice."[63] Without financial support from the union, however, some miners began to trickle back to work in late June. Then on 2 July, faced with that fact that Chinese labourers were now working in No. 1 mine and a second shift was about to start work, Ladysmith's unionized miners voted by a majority of 168 to 117 to accept Dunsmuir's proposal and return to work.[64] Although reporters believed that this would end the strike at Cumberland, the miners there held out until 21 August.

The strike was an unambiguous defeat for the miners. Not only had their attempt to unionize failed but shortly afterwards the WFM ceased to operate on Vancouver Island. Over the course of the next few months, the UMWA replaced the WFM in the interior mines. Moreover, the miners who returned to work were forced to accept worse contracts and lower wages than before the strike. The failure of the strike and the collapse of the union did not spell the end of working-class mobilization in the Island's coal-mining communities, however. On the contrary, the mobilization of the miners entered a new political phase with their endorsement of socialist candidates in the provincial election of October 1903. In effect, class conflict had moved from the economic to the political arena.

The 1903 vote, which came on the heels of the strike wave, resulted in the election of socialists James H. Hawthornthwaite in Nanaimo and Parker Williams in Newcastle, and marked the culmination of a period of working-class political consolidation and mobilization. Until 1903, the political left had been unable to create a unified political organization to represent the interests of the working class. The early socialist movement – beginning in 1901 with the formation of the Socialist Party of British Columbia (SPBC), which three years later merged with the Canadian Socialist League to create the Socialist Party of Canada – was divided into two distinct camps, the moderate reformers, most of whom were active trade unionists, and the "impossiblists," a vocal minority enamoured with Marxist philosophy and preoccupied with theoretical issues. Perhaps best embodied by E.T. Kingsley, editor of the *Western Clarion*, the official newspaper of the SPBC, the impossiblists denied the efficacy of gradual reform and hence distanced themselves from immediate, practical changes that would have improved the conditions of the working class. They argued instead that the only way to empower the workers and improve their lot was through the overthrow of the capitalist system. One consequence of this division was the proliferation of a number of short-lived political parties, such as the Revolutionary Socialist Party of Canada, which consisted primarily of Nanaimo-area miners disaffected with the SPBC, and the Provincial Progressive Party (PPP), a reformist organization based in the BC interior.[65]

Despite the lack of unity on the political left, the labour movement rapidly and fundamentally changed British Columbia's political landscape. In late 1900, Ralph Smith, the moderate Liberal-Labour MLA from Nanaimo, resigned his seat to run as a candidate in the federal election. He was replaced by acclamation by Hawthornthwaite, who, despite close ties to Smith and the Nanaimo reform movement, was moving in the opposite direction politically. Originally a candidate of the Independent Labour Party, in 1902 Hawthornthwaite joined the short-lived Revolutionary Socialist Party when the Nanaimo miners split from the SPBC, before running as an SPBC candidate in the 1903 election. The frequent changes of political affiliation did not harm his political prospects. In 1903 Hawthornthwaite was accompanied to Victoria by the Welsh miner Parker Williams, who won 40 percent of the vote running in the newly formed riding of Newcastle, which included Ladysmith and Extension. Together with the labour candidate from the Kootenays, William Davidson, these men held the balance of power in the new minority government of Premier Richard McBride.

The success of the socialists in 1903 has been much discussed in the literature. Many have interpreted it as the triumph of the SPBC's revolutionary socialism or "impossiblism" over the forces of moderate reform. In reality, however, the distinction between impossiblists and moderates was not especially clear, which suggests that the very terms "radical," "revolutionary," and "impossiblist" obscure more than they illuminate. Both the left wing and the right wing of the movement defined socialism as meaning "that all the mines, workshops, factories and other means of production should belong to the producers," and both rejected violence as the means of attaining this goal, advocating instead the need to work within the parliamentary system. While both sides believed, like Karl Marx, that the revolution was inevitable, socialists in British Columbia were content to wait patiently for it to arrive.[66]

This is clearly evident in the speeches of Hawthornthwaite and Williams. At the first major rally of the 1903 election campaign in Nanaimo, held before an enthusiastic crowd of four to five hundred men and women, Williams and Hawthornthwaite were the keynote speakers. Responding to questions about his recent defection from the moribund Independent Labour Party, Hawthornthwaite expounded at length on the socialists' program. While Hawthornthwaite believed that "Labor [has] completed all its evolutionary stages, but so has capitalism, and the end is at hand,"[67] he never suggested that the miners man the barricades. They neither "promoted revolution," as McCormack has claimed, nor did they accommodate "their impossiblism to political realities" in order to win the election.[68] Like their German social democratic contemporaries, the two Vancouver Island socialist MLAs can perhaps best be described as "evolutionary," rather than "revolutionary," socialists.[69] The main weapon in the socialists' arsenal, Hawthornthwaite proclaimed, was the ballot, the "Gatling gun of political power." Once in power, they would proceed with "the conversion of the means of wealth production into the property of the commonwealth by legal enactment as rapidly as possible."[70] Even then, the socialization of industry would proceed peacefully and legally: they "could not and would not attempt to rudely establish Socialism."[71] While Kingsley wavered on the issue of compensation for capitalists, for example, Hawthornthwaite was adamant that even the most extreme socialist would grant the former owners of the means of production "compensation such as would satisfy the capitalists."[72] And in a remarkable concession, he admitted that even if "British Columbia establishes a Socialist government before Great Britain, the United States and Germany, it would have to wait the triumph of the movement in those countries."[73]

Comforted in the conviction that the revolution was inevitable and imminent, Hawthornthwaite was content to agitate for immediate legislative reforms to ameliorate the condition of the working class in British Columbia. He was proud of his record in Victoria. He had been "sent to agitate for the rights of the worker," he declared. "He had agitated for them, and the moment he ceased to agitate it was time for them to turn him down." And although he believed that trade unionism "could do no permanent good," he argued that it was crucial for the worker to unionize because unions "provided a fighting ground for him where he and his fellows could think out and work out their problems."[74]

The two socialist candidates had different styles and approaches to the campaign and the pressing political issues of the day, a reflection undoubtedly of their backgrounds and experiences. Hawthornthwaite, dubbed the "socialist intellectual" by historian Allen Seager,[75] emphasized the theoretical aspects of contemporary socialist doctrine, although he was by no means doctrinaire. An Irishman who had arrived in the province in 1885, he, like so many other prominent socialist figures in North America and Europe, came from the middle class. He worked initially as secretary to the American consul in Victoria and then as a clerk for the VCMLC in Nanaimo. However, he was apparently dismissed by Samuel Robins in 1902 because of his political attacks on Ralph Smith's Lib-Lab position, after which time he became a professional politician who also later speculated in real estate.[76]

Hawthornthwaite's approach did not alienate him from the mining population; his speeches earned him great applause at public meetings and his percentage of the popular vote in the coal-mining riding of Nanaimo City increased steadily, from 44 percent in the 1903 election to 50 percent in 1907, and to an impressive 63 percent in 1909. Still, the contrast with the Welshman Parker Williams could not have been starker. Unlike Hawthornthwaite, Williams was a coal miner who had left his native Glamorganshire, South Wales, for Canada in 1892. After literally working his way westward – in logging, railway, and coal-mining camps – he and his family finally settled in the Cedar district in 1898. A practical miner by trade, he soon became a practical politician by profession, winning the Newcastle riding in four consecutive elections, beginning with his narrow victory of 1903. Whereas Hawthornthwaite held out the promise of a future socialist state dominated by the workers, Williams concentrated on more immediate concerns, such as wages, safety, and the right to organize. He studiously avoided discussing the finer points of socialist ideology, deferring perhaps to Hawthornthwaite, as for instance when he announced without further elaboration during the 1903 campaign: "The Socialist party's platform was always there, and was the same at all other times as at election times."[77]

In 1903 Williams had to compete with popular Conservative and Liberal candidates in the Newcastle riding, which included Cedar, Northfield, and Ladysmith, and won the seat only because the opposing candidates split the vote. In the predominantly rural district of Cedar, he came in third, while in Ladysmith, with its large population of miners, he received only marginally more votes than the Conservative candidate, Andrew Bryden, the manager of the Extension colliery. His only majority came in the small district of Northfield, which, like Ladysmith, was a mining community. Williams also had to contend with an image problem. He was decidedly unpolished and crumpled in appearance. Ten years after his first election campaign, one writer would declare: "No amount of grooming and combing would transform Parker Williams into a Beau Brummell." While his clothes, "thrown on him in a hit and miss fashion," his unpolished boots, and his callused hands "may have caused some mirth to his colleagues at the capital," his passionate defence of the rights of his constituents, the coal miners and their families, earned him their lasting respect and admiration.[78]

His reputation among the middle class was not as glowing, however. Questions of personal appearance aside, some considered him to be a "fire-eating ogre" and "fanatic," "who would not hesitate to precipitate the nation into a civil war which would make the horrors of the French Revolution pale into insignificance in comparison, and one who would without mercy, banish all home life and have mankind revert to the barbarian state of the dawn of civilization."[79] While not quite as extreme, the businessmen and shopkeepers in Ladysmith did not regard him with any more respect. Although the *Ladysmith Chronicle,* the mouthpiece of the local business community, was "without sympathy for Mr. Hawthornthwaite's methods," it nonetheless recognized his ability: "he has brains, and he knows how to express himself, two factors in which his colleague from Newcastle is woefully lacking." Williams, the newspaper concluded, was nothing less than a "mouthing mountebank" who must inevitably "pay the penalty of mediocrity."[80]

Scholars have questioned why the socialists enjoyed such success in the coalfields of Vancouver Island. Over the years, the responses have become clichéd. The geographic isolation of the miners, the British radical tradition, the demographic structure of the communities – in short, the standard repertoire of Western Exceptionalism – have all been cited as possible causes of the miners' "militancy" and "radicalism." But while the "spittoon philosophers" may have genuinely regarded themselves as revolutionaries, the miners, as Belshaw has argued, generally speaking did not.[81] Testimony given at the Royal Commission of 1903 suggests that miners' commitment to socialism, whether revolutionary or reformist, was at best lukewarm, and the authorities' attempt to connect union activism with socialist agitation was not successful.

Belshaw detects a number of factors that contributed to the mobilization of the mining communities, including nationalism, anti-Asian sentiment, and the desire to protect their rights as contractors of labour and as "penny capitalists" against the rapacious monopolists, but not Marxist theory. Thus, the veteran Nanaimo miner Arthur Spencer, who had been a member of the Knights of Labor in the 1880s, denied that his decision to join the WFM implied acceptance of its socialist political program. The Scottish miner and union treasurer at Cumberland, David Halliday, claimed that he was motivated by his desire to exercise his rights "as a freeborn British subject." For Halliday, the issue at stake was the right to free speech and independent political opinions.[82] The reason why miners endorsed the WFM was neatly summed up by Samuel Mottishaw Sr., when he declared that "a union by itself [i.e., a regional union as opposed to an international union] has no backbone. We did not want a useless organization."[83]

If socialism was not the main motivating force behind the decision to support the WFM, why, beginning in 1903, did the miners embrace the socialist candidates in the provincial elections? Instead of viewing their support for Hawthornthwaite and Williams as a form of deviant behaviour, we should perhaps look at the positive appeal of these men. Simply put, during a time of economic uncertainty and political change, the socialists represented and fought for the immediate interests of the miners and the working class, whereas other candidates manifestly did not. Once again, Belshaw provides the most convincing account of the miners' shift to the left. According to Belshaw, there had been a "steady polarization of interests on the coalfield" during the last fifteen years of the nineteenth century.[84] Given the relative labour peace that prevailed between 1891 and 1903, and the constructive labour relations practised by Sam Robins in Nanaimo, Belshaw suggests that the decisive factor in this process of polarization was the aggressive, confrontational behaviour of James Dunsmuir.

Whereas previously miners had perceived a community of interest between themselves and the operators – as is evident in their support for management candidates in provincial elections – by the early years of the new century this was no longer the case. Although James Dunsmuir was elected to the legislature in South Nanaimo in 1900 and subsequently became premier, few if any measures taken by the government during his tenure benefited those who elected him to political power. On the contrary, against the background of anti-labour legislation, the mobilization of hard-rock miners in the interior of the province, a mine explosion at Cumberland, and a scandal involving the sale of the E&N Railway,

Dunsmuir succeeded only in alienating his mining constituents. According to Belshaw, "it was at this stage – when the remaining distinctions between the employer and the state were virtually obliterated – that socialism (as an alternative political and economic system) became attractive."[85]

As important as the deteriorating relationship between Dunsmuir and his workers undoubtedly was – and regardless of how attractive socialism as an alternative political and economic system became – the polarization of British Columbia society must also be seen in the context of broader economic and political developments. Throughout the industrialized world between about the 1890s and the outbreak of war in 1914, a period often referred to as the "second industrial revolution," fundamental changes occurred in the way the capitalist economy functioned.[86] On a worldwide scale, capitalism was gradually transformed from a system of essentially small and medium firms, family businesses, and paternalistic entrepreneurial production into a predatory system dominated by highly concentrated forms of corporate production. The enormous increase in corporate mergers, dependent upon large-scale financial and capital commitments, created huge commercial and financial empires that radically altered the nature of economic production.[87] In order to improve efficiency, production, and competitiveness, and to check the growing influence of trade unions, capital placed increased pressure on wages, introduced new technologies, and sought to exert greater control over the workplace through new forms of systematic, "scientific" management.

These developments had a tremendous impact on the lives of the working class; they represented nothing less than a frontal assault on the traditions and status of skilled labour.[88] Skilled workers and craftsmen in the industries most affected by this process of rationalization and bureaucratization were removed from production through deskilling and proletarianization and transformed into unskilled operatives, while at the same time a new class of supervisory personnel was created.[89] As a consequence of this "revolution in the workplace,"[90] working-class attitudes hardened, unions expanded, and labour unrest intensified, especially among skilled and young semi-skilled workers in places where attempts to rationalize production were successful and systematized. Moreover, the nature of strikes also changed, as more and more they focused on issues of power and authority, challenging not only management's attempt to erode worker control of the workplace but also the position of the working class within the general social and political order.[91]

We must also understand the trend towards the formation of employers' associations within this context, as part of an aggressive campaign against working-class mobilization, which greatly contributed to the polarization of the working class and employers.[92] Employers' associations were designed to fight labour and to lobby government for favourable legislation. Among the earliest employer associations in BC were the BC Shingle Manufacturers' Association (1894) and the BC Salmon Packers' Association (1898). Over the next few years more were formed, representing a variety of industries. The most important was the Employers' Association of Vancouver, an umbrella group that sought to protect the economic and political interests of business and to shape public opinion against the labour movement. The major catalyst was the UBRE strike of 1903, which convinced employees that a united front against the trade union movement was the only way by which they could re-establish "control over the labour market and workplace."[93] By 1905, British Columbia had twenty-six employers' associations, whose stated goal was to retain control over the workplace and "to return managerial power to owners and representatives of capital."[94]

Andrew Yarmie suggests that employers' associations bore considerable responsibility for the increase in class conflict because they adopted a framework for industrial relations based on coercion and conflict rather than negotiation. For this, there was a long precedent in the Vancouver Island coal-mining industry, where the Dunsmuirs had employed such tactics as early as 1877. In this environment hostile to labour, employers had always found a willing and compliant ally in the state and government. Rather than being a neutral mediator between capital and labour, the state became the guarantor of capitalist production and the underwriter of existing social and economic relationships.[95] But this did not satisfy the business community. When government tried to (re-)legitimize itself or stabilize the social order through the enactment of various reforms, the employers' associations attacked state intervention as inimical to the interests of free enterprise, even though, Yarmie concludes, the concessions made by the state to popular interests "did little to alter the existing power relations in society."[96]

The polarization of economic interests – the workers' gradual realization that their interests were not the same as their employers' but were indeed inimical – had a direct impact on the political culture of the province. The rise of working-class political parties was a natural consequence of the inadequacies of the existing political structure in British Columbia. The system of notable politics or personal government, whereby "political alignments [were] based on tenuous personal pledges"[97] rather than by party affiliation, had entered a period of profound crisis during the last years of the nineteenth century with the succession of five ineffective and corrupt administrations between 1898 and 1903. This system of notable politics had singularly failed to address the concerns of the province's growing working-class population.

But even with the advent of party politics after the appointment of McBride as premier and during the subsequent election campaigns of 1903 and later – some historians have argued that McBride adopted party lines in order to combat more effectively the labour and socialist vote – the Conservative and Liberal political parties were unable to attract the support of the miners.[98] This failure is perhaps even more remarkable given the inherent difficulties the workers had in developing an independent party, their divisions along socialist, reformist, and Lib-Lab lines, and cultural factors such as race and sexism, all of which undermined class solidarity.[99] The reverse of the coin, however, was that the failure of the middle-class parties to include the workers resulted in the emancipation of the working class from bourgeois political tutelage. In other words, excluded from political participation first of all by the political cliques of traditional notability that had dominated political life in British Columbia up to 1903 and then from the new party regime imposed by McBride, workers were forced to adopt a new style and language of politics in order to enter the political sphere in any meaningful way. Thus, they rejected not only the self-interested paternalism of Dunsmuir but also the moderate Lib-Labism of the popular Ralph Smith.[100]

As a consequence, the rise of socialist politics was less a reflection of a newly discovered radicalism and militancy than of a new political culture that was polarized along class lines. By 1903, political culture in British Columbia had been refashioned. Population growth, changes in the economic structure, the rise of the interventionist state, and the inadequacy of personal government had realigned political discourse in the province. The growing support given to working-class parties and the rise of trade unions heralded the beginning of a new style of politics in British Columbia, as was indeed the case throughout the industrialized world. A new pattern of partisan political culture, based on the clear articulation of group interests, produced new politicians, policies, organizations, and conflicts. This was a period of popular mobilization that challenged the older hierarchies of subordination and redefined identity in increasingly populist, racial, gendered, and class terms. Gone were the days of politics by notables, whereby prominent, respected individuals were acclaimed into power by virtue of their social rank and position and without grassroots support. Symbolized best by the emergence of modern political parties during the 1903 provincial campaign, British Columbia had entered the age of modern politics.

# 6

# *The Great Strike*

The failure of the 1903 strike was a major setback in the drive to unionize the Vancouver Island mines and for the provincial labour movement generally, but this did not end the mobilization and organization of the coal miners. Often acting in secret, the miners saw their efforts enhanced in the first decade of the century by the general buoyancy of the local economy, the scarcity of skilled labour, and the lack of improvement in working conditions underground, a fact that was brought home to them once more in the wake of the 1909 explosion at Extension. The arrival of the United Mine Workers of America (UMWA) in late 1910, the culmination of years of dedicated work by union activists, set the stage for inevitable conflict. Both the coal companies and the provincial government were alarmed by the presence of the UMWA and were determined to stop it in its tracks. From their perspective, organized labour was the implacable foe of capitalist production, to be fought vigorously at every opportunity, with every available means. The confrontation came in 1912 and lasted two long, difficult years. Although the unity and determination of the strikers was impressive, even with the union's massive support they could not withstand the combined onslaught of the companies and the state. When the strike collapsed in the summer of 1914 after two years of sacrifice and hostility, the miners' defeat was total, leaving the communities bitterly divided and effectively ending the drive to unionize the Island mines.

In a lot of ways, the Great Strike was like previous strikes in the Island mines. Traditional grievances, such as wages, working conditions, and safety, played a role. Ultimately, however, as in 1903 this was what historians refer to as a control strike.[1] At issue were the miners' rights, challenged by management, to exercise basic control over the organization of the workplace, including the hiring and firing of employees. The struggle for union recognition was also as much a question of control as it was an economic concern, for unions would empower the men where it counted most: at the working face. Where this strike differed from others was in the relative strength of the miners on the one hand and the power and determination of the companies and the state on the other.

Backed at last by a powerful union – and one with very deep pockets – the majority of the mining population endorsed the actions of the UMWA and during the dispute demonstrated remarkable solidarity, perseverance, and endurance in the face of an overwhelmingly hostile coalition of forces. Despite persecution by the authorities, mass arrests, jail terms, and occupation of the strike zones by the provincial militia, most strikers remained steadfast in their aims, confident that they would win this high-stakes battle. The odds were stacked against them, however. Although their victory was by no means a foregone conclusion, the companies, especially Canadian Collieries (Dunsmuir), Limited, or CC(D), which had bought out Dunsmuir's holdings in 1910, enjoyed "unusually favourable advantages for resisting the strike."[2] This was not only the result of its substantial financial resources and political clout; an equally decisive factor was the fact that the strike occurred just as the economy began to slip into recession. In the long run, the company had little economic incentive to settle the dispute and was able to wait out the striking miners, content to produce what they could with the available strikebreakers.

The longest labour dispute in the history of the Island coal industry began inauspiciously enough. In September 1912, Oscar Mottishaw, a practical miner, was dismissed from his job in the Union mines at Cumberland, where he had recently been hired under contract by a miner named Coe. According to mine superintendent R. Henderson, Mottishaw, who had been on the job for only three days, was fired because Coe was paying him $3.50 per day rather than $2.86 per day, the stipulated company wage for helpers. This was not the first time that Mottishaw had been dismissed from Canadian Collieries, however. Mottishaw was a union organizer for the UMWA, and until August 1912 had been working at Extension, where he and Isaac Portrey were members of the gas committee. On 15 June, they had filed an official report documenting the presence of gas in five places of No. 2 mine. The Ministry of Mines inspected the mine for gas and found a small quantity in a caved-in portion of No. 3 stall. This was enough for the inspector to order the company to improve ventilation in the affected district, to ensure that miners used safety lamps, and to reduce the number of men working in the area. A few weeks later, the inspector reported that the company was complying with his orders.[3]

By all accounts the matter should have ended here. The gas committee, the provincial mine inspectors, and the company had satisfactorily fulfilled their obligations under the Coal Mines Regulation Act (CMRA). However, the sale of Dunsmuir's Wellington Colliery Company to CC(D) two years earlier had not changed its policy of firing union activists. When Mottishaw's place ran out in August, he was refused another, which was tantamount to dismissal. Even though he applied for other work, the management of Extension refused to take him on again, and instead hired other men.[4] As a result, Mottishaw was forced to look elsewhere for work, and eventually left Extension for Cumberland. Although CC(D) officials adamantly denied that they were discriminating against Mottishaw – they always maintained "the right to hire and discharge unquestioned"[5] – he had clearly been targeted by the company's management. As a union man, he was considered by officials to be a serious "trouble-maker and agitator" who not only interfered with the firemen but used abusive language – he apparently called them "scabs" and "blacklegs" – and even attempted to intimidate his coworkers.[6]

In response to Mottishaw's dismissal from Cumberland, the miners held the first of a series of meetings on 15 September and, according to union representatives, decided "to take a holiday as a protest against the action of the management of the company in discriminating against two of its employees."[7] Later that evening, union members held another meeting, during which they endorsed the previous recommendations and appointed a committee to discuss the matter with management. On 16 September, the day of the declared "holiday," the committee, accompanied by Robert Foster, president of the UMWA local District 28, went to speak with the management of CC(D) but was refused an interview. When the miners returned to work on the morning of 17 September, they discovered that they had been locked out. The company had posted notices at the mine ordering the men to remove their tools and informing them that they would be paid off. It also notified the men that if they wanted to work for CC(D) again, they would have to sign a two-year yellow-dog contract under the old conditions.[8]

In the wake of the company's unexpected response, the miners attempted yet again to meet with management in the hope of peacefully and promptly resolving the growing dispute. Once more the company refused to speak with the committee. The company cashier, G.W. Clinton, derisively referred to by Foster as the "would-be czar of Cumberland," summed up the company's attitude when he declared: "We don't want to hear you at all."[9] This refusal to negotiate an end to an essentially internal dispute radically changed the complexion of the conflict and made confrontation inevitable. The men, not anticipating a lockout, had called a "holiday" rather than a strike, because this was the only way all three of the shifts could get together. The protest was intended to last one or two days at most, at which time the men, having made their point, would return to work.[10] This was by no means an unusual strategy. It was not uncommon for the miners to stop work for funerals, important holidays, or even occasional protests against management. Instead, faced with an intractable company, the miners now had to respond to the changed circumstances, which they did by escalating their demands. They now "decided to take a holiday until the management would meet the committee and provide some means for the adjustment of difficulties and differences that may and do arise around large industrial operations."[11] When news of the events in Cumberland reached Ladysmith on 18 September, the Extension miners, by a majority of 100 votes, decided to stop work until either Mottishaw was reinstated or work at Cumberland resumed.[12] They too were notified by the company to remove their tools. Almost 1,600 men were now off the job.

The dismissal of Mottishaw set off a chain of events and a flurry of telegrams and activity. Coal strikes, perhaps more than any other form of industrial action, tended to cause panic among the authorities. This was not necessarily because miners were so radical and militant, although the state and public opinion thought this was the case, but because coal mining was considered to be an "essential service." A strike in the coal industry could have a detrimental impact on the general economy. As a result, all levels of government had an interest, albeit varying in degree, in seeing the strike settled.

The first attempt to resolve the dispute was made by local officials led by Cumberland's mayor and the board of trade. Municipal authorities clearly feared the impact of the strike on businesses and the community, but had little sympathy for the demands of the miners. Although they approached both management and the strikers in an attempt to bring the two sides to the bargaining table, and even enlisted the help of a union organizer to negotiate with the company, their efforts to mediate the dispute failed. Colliery management refused to deal with the union committee, and the miners were suspicious of the local businessmen, who told them "how 'prosperous' they were." Annoyed by their arrogance, the miners "showed nothing but contempt for the good offices of the board of trade."[13] Eventually, Mayor McLeod and ex-mayor of Cumberland D.S. McDonald abandoned their attempts to mediate and actively tried to "get outsiders to come in and vote against the strike."[14]

Efforts by the federal government to resolve the dispute were just as futile. Shortly after the strike began, news reached Ottawa that two of the most important collieries on the Island had ceased working. The federal minister of labour, T.W. Crothers, informed the union that under the Industrial Disputes Investigation Act (IDIA) of 1907, the strike was illegal. Because the coal industry was considered a "public utility," it was against the law to declare a strike until an official report had been filed by a board of conciliation and arbitration.[15] For the act to come into effect, however, either the miners or the company had to lodge an official complaint to begin proceedings. This neither side was prepared to do. John McAllister, secretary of local 28 of the UMWA, replied in a telegram on 20 September that the miners had declared a holiday, not a strike, and had been locked out upon returning to work. To apply for a board of arbitration under the act would have been to admit that they had declared an illegal strike.

Furthermore, given the history of IDIA settlements, there was little chance that the board would decide in the union's favour. The act did not protect workers, but instead gave operators the opportunity to undermine their position. As McKay has argued: "Ideologically, the IDIA was a heaven-sent weapon for capital. Manipulating the near-universal respect for the law, which the non-violent character of the labour movement reveals so well, the Act cast a suspicion of illegality and unreason over the pursuit of collective rights."[16] Frank Farrington, the highest-ranking member of the union's international board, later declared that the union would have been willing to apply for a conciliation board under the act if the company did the same.[17] However, CC(D) was not interested in conciliation if it meant union recognition, and discouraged all attempts by the federal government to intervene in the strike. The company apparently believed that the actions of the Island miners would not receive the support of the international union and that the miners would then appeal to Ottawa for a resolution to the dispute.[18] In the meantime, the company was determined to use more traditional methods to defeat the strike and the union.

Of far greater consequence was the attitude of the provincial government. While the municipal and federal governments attempted to resolve the dispute, the provincial authorities were determined not to get involved. On 18 September, union officials J.J. Taylor and Duncan MacKenzie forwarded a resolution to Richard McBride, premier and minister of mines, requesting him "to at once investigate the action taken by the officials of the Canadian Collieries Company Limited, in discriminating against employees acting on Gas Committees."[19] The following day, the miners received their answer: due to lack of information about the discrimination, his department was "at a loss to understand what it is really expected to investigate." Furthermore, McBride continued, the department "does not now propose to be drawn into any labour disputes between employers and employees."[20] Despite sending him further information about the Mottishaw case on 21 September, the union could not persuade McBride to abandon his "unbiased" and "neutral position" and to intervene in order to resolve the dispute. Nor would he investigate charges that CC(D) management had violated the CMRA by interfering with the gas committee. When a delegation of miners headed by Nanaimo MLA Jack Place attempted to meet with McBride a few days later, they were likewise refused a hearing.[21]

McBride was far from neutral. His sympathies were clearly with big business. He had very close connections with the railway magnates William Mackenzie and Donald Mann, owners of Canadian Collieries (Dunsmuir) and the chief force behind the Canadian Northern Railway.[22] His attitude towards the striking miners was one of contempt. According to one source, he thought it was "intolerable" that "coal miners should think they had the right to make demands on the mine owners."[23]

Subsequent steps taken by his regime intensified the conflict. The unrestrained use of poorly trained, aggressive special constables, for instance, who arrived in considerable number at Cumberland in November 1912 and who were later employed throughout the strike zones, greatly antagonized the miners and was interpreted as a declaration of war against the strikers. The government's hard line was reinforced in early 1913, when a resolution introduced by Parker Williams and Jack Place calling

"Lashed to the Mast," *Vancouver Sun,* 8 February 1913. A commentary on Premier McBride's refusal to launch an investigation into the miners' dispute.

for the appointment of a select committee to investigate the strike was defeated in the legislature by a vote of 35 to 2, but only after McBride was forced to quell dissension within the ranks of his own caucus.[24] After condemning the striking miners, McBride justified once again his policy of "neutrality" while at the same time absolving his government of all responsibility for a settlement. A mechanism for resolving labour disputes already existed under the IDIA, the premier declared, and consequently this problem fell within federal, not provincial, jurisdiction. The government did not stand idly by. Although it had initially refused to investigate the strike and its causes, it did agree, under pressure from Vancouver MLA C.E. Tisdall, to launch a commission of inquiry into the shortage of coal on the mainland and its high price to consumers.[25]

This revealed not just the government's stance, but the extent to which the political landscape in British Columbia had changed over the last few years. McBride enjoyed widespread popularity, possessed a large majority in the legislature, and was presiding over a booming provincial economy. He was at the peak of his power. The influence of Williams and Place, however, had declined. Since the election of 1907, they no longer held the balance of power in the legislature, and although they continued to agitate for the miners, speaking at public meetings and rallies throughout the strike zone, they could not persuade McBride to help resolve the dispute. Not only were they effectively ignored by the government but they could also be removed and silenced, temporarily at least, which happened following the August unrest when Place and Williams's son were arrested.

McBride had no political incentive to settle the strike, but his determination to pursue a policy of intransigent nonintervention set the tone of the dispute. By not investigating the charges of discrimination and interference with the gas committee, McBride had effectively denied their validity. By refusing to mediate the dispute, he gave the company a green light to pursue whatever strategy was required in its confrontation with the miners, in the hope that it would result in the defeat of the strike and the union. The state had entered into an alliance with CC(D), which it supported with contingents of provincial police and special constables, whose purpose was to protect strikebreakers and company property. The significance of the provincial government's refusal to mediate the dispute cannot be overemphasized. Not only did it encourage both sides to dig in their heels, directly contributing to the escalation of the labour dispute and the subsequent hostility, bitterness, and violence, but it also guaranteed that this would not be a short strike or one that could be peaceably resolved.[26]

The original protest against the company's discrimination against union men soon became a battle for union recognition. The attempt to unionize the miners on Vancouver Island had met with only mixed success. The Miners' and Mine Labourers' Protective Association (MMLPA) had been successful in Nanaimo during the 1890s, but had failed to win recognition in Cumberland and Wellington under the Dunsmuirs. The collapse of the 1903 strike and the Western Federation of Miners (WFM)

had been a serious blow to the miners. Without the backing of a strong union and fearing company retribution, miners were forced to recruit in secret, and progress was slow. By 1910, however, the circumstances had changed. The 1909 explosion at Extension had been a vicious reminder that working conditions and mine safety had not improved. In addition, the sale of Dunsmuir's interests to Mackenzie and Mann might have encouraged some to think that a new era of industrial relations was dawning. Be that as it may, in late 1910 a committee of union activists approached both the WFM and the UMWA in an attempt to win support. By this time, with the WFM organizing only hard-rock miners, the UMWA was the only viable option for the local men.

The UMWA had been established through the merger of the coal miners' assembly of the Knights of Labor and the National Federation of Miners in Columbus, Ohio, in January 1890. An industrial union that organized everyone who worked in coal mining, the UMWA quickly became one of the most powerful unions within the American Federation of Labor (AFL), with which it was affiliated shortly after its creation. It also became one of the largest unions in the United States; within the space of fifteen years its membership skyrocketed from 10,000 in 1897 to over 353,000 in 1912, with the bulk of its members in Illinois, Pennsylvania, and the US Midwest.[27] In April 1903, the American organizer Harry Bousfield began the first attempt by the UMWA to recruit Canadian miners. Arriving in the Crow's Nest coalfield just as the WFM strike ended in 1903 and around the time of the Frank disaster of 29 April, when a landslide swept down Turtle Mountain, wiping out the town of Frank and the works of the Canadian-American Coal Company, Bousfield and the UMWA were quickly able to absorb existing locals of the WFM. On 5 May 1903, the first UMWA local was formed in Fernie, and was soon followed by locals in Michel, the Morrissey and Tonkin districts, and Coal Creek in the Crow's Nest Pass. In early November, District 18 was formed with Frank H. Sherman as president. In the new year, Sherman began to organize in Alberta, where he met with considerable success. District 18, which included the Kootenays and Alberta, eventually consisted of twenty-seven locals and over 3,600 members.[28]

The early success of the UMWA in eastern British Columbia, Alberta, and later Nova Scotia was not repeated on Vancouver Island.[29] An early attempt to organize miners in Nanaimo in 1905 failed, and for the next few years the executive in the US was reluctant to move into the coalfields of Vancouver Island.[30] Unlike in Nova Scotia, where there was a strong provincial union, and the East Kootenays, where the WFM had been firmly entrenched, the lack of an indigenous union on the Island suggested to the union leadership that interest was not that strong. Accordingly, union president T.L. Lewis informed the local miners that the UMWA would consider organizing them only "once the men on the Island made some show for themselves."[31]

Faced with this lukewarm response by the UMWA, Ladysmith miners took the initiative and formed their own union, known as the Canadian Federation of Miners, in order to demonstrate to the American organization that union spirit was strong on Vancouver Island. The miners successfully recruited members, forming locals in Cumberland, South Wellington, and Nanaimo. Operators were bitterly opposed to this new union, and its members knew that they stood little chance of success against the collective power of the companies. Under pressure from Cumberland miners, the union men once more approached the UMWA. The response from the new president, J.P. White, was much more positive for the miners.[32] He promised to investigate conditions on the Island and to ascertain whether there was enough support for the UMWA. The organizational efforts of George Pettigrew, a Ladysmith miner, bore fruit, and the numerous meetings in the mining communities were very encouraging for the activists. According to one report, 300 men at Cumberland, 400 at Ladysmith, and 170 at the smaller South Wellington operations indicated their willingness to join the American union. The UMWA board agreed to grant a charter, and at a convention held at the opera house in Nanaimo on 5 December 1911, UMWA District 28 was formed.[33]

The state and companies blamed the UMWA for causing discontent in the Island coalfields and for provoking and then prolonging the strike. Both company management and the official inquest into the strike agreed that "not very long after its [the UMWA's] establishment [on Vancouver Island] friction seems to have commenced at these places between the men and the company."[34] As the history of labour unrest in the Vancouver Island coalfields attests, however, friction between the men and the coal operators had existed long before the arrival of the UMWA. High death rates, poor working conditions, and hostile relations between management and men were facts of life in Wellington, Cumberland, and Ladysmith-Extension. The miners certainly did not need a union in order to feel aggrieved by the companies.[35]

These and other complaints have been described as "minor grievances" by one writer.[36] However, given the horrific death rates in the Island mines and the fact that it was extremely difficult for miners to earn a living wage, issues affecting safety and earnings were never minor. What was different in 1912 was the presence of the UMWA on the Island, which acted as a catalyst for the miners. With its large membership base, its strong financial backing, and its impressive organizational commitment, the men became increasingly confident that the international union would help redress these wrongs and end the company's tyrannical hold over them. The miners' faith in the power of the union was reflected in their efforts to organize in the months prior to the strike and in the fact that they were no longer reluctant to air their grievances or demand improved work conditions. The presence of the union empowered the men, and during the summer of 1912, as the union grew stronger, the miners' demands for positive change in the workplace intensified. The *BC Federationist* reported just weeks before the strike that "rumblings of the coming upheaval, unless conditions are ended or amended, are becoming more pronounced from day to day." The situation in Cumberland was especially serious: "trouble was brewing in that territory."[37]

Equally significant, the union acted as a catalyst for company management. As support for the UMWA grew, local companies, led by CC(D), intensified their efforts to stop the union in its tracks. From the very beginning, the colliery operators pursued a highly antagonistic, confrontational policy towards the union. CC(D) in particular was determined to crush the UMWA. As the union tried to organize, the company embarked upon a campaign of harassment and intimidation against union members and organizers. This campaign was conducted not just by the coal-mining companies – this was, of course, to be expected. Local authorities, evidently acting at the behest of the colliery operators, also tried to intimidate union members by placing petty obstacles in the way of the miners attempting to organize. Under pressure from CC(D), Cumberland city council tried to prevent the union from collecting union dues inside the local bank where the men were paid, arguing that it had not received special permission. In August 1912, public street meetings were prohibited in Nanaimo and the public bandstand had been removed because it was used to hold union meetings. All of this was proof, commented the *BC Federationist,* that "the coal companies are getting alarmed at the spread of organization among the men."[38]

Just as the UMWA did not manufacture imaginary grievances, it was not a monolithic entity. Indeed, the strength of the UMWA, while not inconsiderable, has been greatly exaggerated. The miners were not simply passive pawns at the mercy of the powerful international union, as the local mine operators, such as James Dunsmuir, liked to think. Official commissions and inquests, contemporary newspaper stories, and even some recent historical studies speak of the nefarious influence of foreign agitators, of unfair union votes and stuffed ballot boxes, and of intimidation in the union halls or in the workplace. But the men were neither ignorant nor gullible; they were not naive victims of unscrupulous manipulators. Instead, they demonstrated an acute awareness of what was and was not politically possible, and were quite capable of formulating their own demands in the volatile political climate.

In many cases, policies and actions were determined by the striking men and their locally elected union officials, not by the international board, led by Frank Farrington, that the union executive in the US had placed in charge of the strike. Initially at least, even local union leadership was rarely in a position to dictate to its members; more often than not, it was forced to react to their demands. Throughout the strike, union officials had to play a frustrating game of catch-up with the striking miners in order to keep abreast of developments and to assert their leadership and authority. This was illustrated at the very beginning. Cumberland miners decided to take a holiday in response to the dismissal of Mottishaw without consulting George Pettigrew or David Irvine, the two senior local union officials. In fact, the decision to strike occurred at a very inopportune moment for the UMWA. It had not completed its organizational efforts – few inroads had been made in Nanaimo, for instance – and the executive did not feel strong enough to fight CC(D). As Wargo has correctly pointed out, the union leaders were forced "to go along with the strike" or risk losing credibility.[39] The fact that the UMWA was relatively new to the Island and that it was trying to garner as much support as possible meant that this was a risk that senior union officials were not prepared to take.

Conflicts between the international board, local union officers, and the membership were evident throughout the strike. One problem was the union's strict policy of nonviolence. The UMWA knew from experience that violence inevitably alienated public opinion and led to retaliation by the state and company, which had the instruments of power and repression at their disposal. As the strike dragged on, union leadership had difficulty restraining its frustrated members. The first serious instance was in Cumberland in November 1912. The situation there had become increasingly tense because production had resumed with the employment of Chinese and other strikebreakers. The striking miners responded by marching and protesting in the streets and by employing the age-old tactic of "rough music" or "charivaris" in an attempt to persuade the strikebreakers to quit work. In response, local authorities requested the provincial attorney general, William Bowser, to send in a contingent of special constables, which only further antagonized the striking miners. As hostilities escalated between the strikers, the strikebreakers, and the special constables, union officials were forced to step in, urging its members to keep the peace, to stop their street demonstrations, and to refrain from antagonizing the strikebreakers.[40]

Just as the situation cooled down in Cumberland, it threatened to become explosive in Ladysmith. The management of Canadian Collieries initially believed that if they could break the strike at Cumberland, then the miners at Ladysmith would also return to work. As a result, the company concentrated its efforts on resuming production at Cumberland and did not attempt to reopen its Extension works until January 1913. As in Cumberland, the attempt to restart the mines and to import strikebreakers mobilized the strikers into action. There were mass demonstrations and strikers blocked sidewalks in a show of strength. Police reported a number of incidents as striking miners began to harass strikebreakers. Two miners, Steve Mrus and Steve Puyanich, were arrested and charged with threatening and intimidating the strikebreaker Joseph Lapsansky, whose house was later targeted during the August unrest. On 31 January, James Nimo, a fire boss at Extension, was "set upon and maltreated by two unknown persons."

Perhaps because he feared a repeat of the more serious disturbances in Cumberland, Ladysmith's mayor panicked and on 28 January 1913 sent a rather alarming telegram to the attorney general in Victoria, declaring: "Situation out of control of city police suggest provincial police take over city and maintaining order."[41] Such extreme measures were hardly necessary. When constable George Hannay met with local union officials and about one hundred striking miners "with a view to have all disorders, demonstrations and streetblocking cease entirely," union officials were more than willing to accommodate him. In fact, Irvine and Pettigrew threatened that if the men broke the law and continued to harass the strikebreakers, the union would withdraw its support for the strike.[42] In the end, the eight additional police officers sent from Victoria were not overly taxed.

Historians have resolved the discrepancy between the actions of the miners and the reaction of the union by describing the decision to take a protest holiday as "impulsive" and "precipitant," a result of the coal miners' radicalism and "blind enthusiasm for strike action."[43] This suggests that the miners, unaware perhaps of the broader consequences of their actions, were behaving irrationally. On the contrary, the decision to strike was never taken lightly but was the result of serious, reasonable, and often lengthy deliberations. Strikes were always a last resort, taken only after all other attempts to mediate disputes had been tried.

In the wake of the lockout, the miners rejected the yellow-dog contract of the company and decided to stay off the job. They clearly believed that under the present conditions they were in a good position to win the struggle. The decision met with overwhelming support; the economy was still buoyant and demand for their product was strong (the economy would enter into recession only after the strike began), and perhaps most important from their perspective, they were supported by a strong international union. Rather than acting irrationally and impulsively, the miners believed that they were acting from a position of strength at a propitious moment. None of this is to suggest that there was a conspiracy to strike between the union and the men. While the UMWA provided the miners with a strong vehicle through which they could collectively express their grievances, fight for improved working conditions and wages, and bring about some form of industrial democracy, it did not call all the shots.

Canadian Collieries proved to be as formidable a foe of the UMWA as James Dunsmuir had been of the WFM. It has been suggested that the new company welcomed the strike because it was unable to pay dividends to its shareholders, who had been issued worthless shares. Lynne Bowen, pointing to the company's decision to recruit strikebreakers and to resume production as quickly as possible, has discounted this view.[44] However, if the company did not exactly desire a strike, the strike nevertheless came at an opportune moment, and CC(D) was more than willing to take advantage of it. Management saw the strike as a chance to destroy the union movement once and for all, and to reduce their costs, especially once the economy slid into recession, all the while maintaining sufficient production by employing the Chinese and strikebreakers.

To combat the union, Canadian Collieries employed the same tactics that had been successfully used by Robert and James Dunsmuir during previous labour disputes. On 27 September 1912, the company began to serve eviction notices to those striking miners who lived in company houses in Cumberland. The men and their families resisted as long as possible, but as the strike dragged on they were eventually forced to abandon their homes. A few days later, CC(D) general manager W.L. Coulson wrote to J.R. Lockhart, superintendent at Cumberland, requesting the names of the worst agitators, because "we may want to take legal action against the instigators."[45] Although Lockhart was prompt in his reply, the company apparently did not proceed with its action.[46]

The company also moved quickly to hire Chinese and Japanese workers as strikebreakers. Initially, the Chinese and Japanese at Cumberland, along with mine engineers and fire bosses, refused to break ranks during early attempts to negotiate with the company, even though they were apparently offered a wage increase of $1 per day.[47] In order to induce the Chinese and Japanese back to work, Canadian Collieries had to use force. On 24 September, G.W. Clinton, accompanied by a dozen or so special constables, sealed off Chinatown from the rest of the population. Once the Chinese were isolated from the striking miners, the police threatened to evict them from their homes if they did not return to work at once. In order to pressure the Chinese mine workers further, the authorities also visited local Chinese merchants and demanded that they cut off credit to the Asian miners.[48] As a result, they eventually coerced between 150 and 200 men into signing the new two-year contract. At the same time, company officials went to the logging camps and woods where Chinese men were employed and brought them back to Cumberland in order to vote on a union motion to return to work. When the strategy of Canadian Collieries became clear, the union responded by simply calling off the vote.[49]

The UMWA certainly welcomed this initial display of solidarity by the Chinese and Japanese workers. Since 1907, the union in the United States had officially opposed exclusion and had even begun recruiting Asian miners.[50] However, UMWA policy towards the Chinese was conflicting and should not be misinterpreted as one of total acceptance of the Chinese as equals. The decision to include or exclude Asian workers was ultimately left up to the individual locals. The constitution of the Fernie local, for instance, refused to admit Chinese and Japanese members.[51] Likewise, on Vancouver Island, union officials did not even attempt to include the Chinese in their organization, largely because they experienced great difficulty overcoming the racism of the European population, even when its racism clearly conflicted with its best interests. In early July 1912, to cite one important case, the union energetically supported an attempt by striking Chinese stokers working at the pithead at Extension to win a wage increase of twelve cents a day, which would have brought their daily rate to $1.75. Much to the embarrassment of the union, Europeans were hired as strikebreakers after the company promised to employ only white stokers in the future. In response, the union claimed that it had "expelled those of their members who have taken the places of the Chinese" and declared that it was "prepared to come out as a body if the Chinese [were] not reinstated at the increased wages they are asking for."[52] The ultimate gesture of the solidarity of the white miners was not put to the test as a settlement was soon reached between the company and the Chinese stokers. As the union and the labour press had predicted, however, the company did not live up to its agreement with the white miners. By August the Chinese strikers were back at work and the white strikebreakers had been dismissed.[53]

Several lessons could have been learned from this incident after the strike began in September 1912. First, the Chinese stokers' strike at Extension once again revealed the limits of the UMWA's authority over the miners. Despite the union's attempt to direct the men and follow a consistent policy, in reality it was obliged to tailor its activities to the demands of its constituents. Notwithstanding expulsions and threats, the UMWA had difficulty proceeding against the collective will of the white miners, which was hostile to Asian labour. A second and perhaps more important lesson was learned by the Chinese. Their experience in this short strike taught them that they had no reason to expect any support from white miners in their struggles against the company. Given their general isolation from the community, the lack of solidarity demonstrated by the European workers, and the naked coercion of the company, it is little wonder that in September the Chinese and Japanese employees chose to return to work rather than continue with the strike.

Canadian Collieries encountered other difficulties at the beginning of the dispute. Unionized railway men refused to handle coal cars that were loaded by non-union men,[54] and engineers and fire bosses walked off the job. Without these men, not only would it have been impossible to resume production but regular maintenance would have ceased and the mines could have flooded and accumulated gas. Like the miners, the engineers and fire bosses had grievances with the company, although few actually endorsed the strikers' position. In August, the engineers had joined the British Columbia Association of Stationary Engineers and sent a committee to discuss wages with management. According to company policy, management declared that they would not meet with union representatives. When fire bosses also formed a union and made similar demands, they were told the same thing, and during the first few weeks of the strike, managers were forced to do the work of the engineers and fire bosses. G.W. Clinton added the jobs of weigh boss and car pusher on the tipple of Union No. 7 to his regular duties as company cashier; some pit bosses were now driving mules and shotlighting; and D.R. McDonald, the superintendent of the railroad, was teaching the Chinese how to fire shots.[55] The outcome of the strike might have been different had the engineers and firemen maintained solidarity with the miners. The executive of the engineers' union, however, was not happy that its members in Cumberland had come out in support of the miners. It condemned the decision in a telegram and urged members to return to work, writing that "by the members taking it upon themselves to get mixed up in the matter without consulting the executive on a matter of such grave importance, the executive is not in a position to support the members of the Cumberland lodge."[56] As a result, when the company offered them a 20 percent increase in wages and even agreed to recognize their unions, they returned to work.[57]

Production quickly resumed at Cumberland, but the striking miners and their families still demonstrated a remarkable degree of solidarity. As the dispute progressed, they were confronted with severe economic hardship, an increasingly hostile and repressive state, and an inflexible company determined to crush them and their union. Although some striking men returned to work, their number was still relatively small, and the strikers' determination to win did not diminish even as their frustrations mounted. No doubt the presence of the UMWA contributed to this. The strike pay was essential, of course, but so was the union's organizational ability and its commitment to the community. The union took numerous steps to maintain solidarity and strengthen morale. In Ladysmith it rented a large 500-seat hall, with six games rooms and a reading room, and even bought a piano, which the people used regularly for concerts. During the first Christmas holidays of the strike, it was a hive of activity. Families decorated it and distributed toys, clothes, sweets, and oranges to the children of the striking miners: they even "had an entertainment of moving pictures." The situation was worse in Cumberland, and Christmas holidays there were not as festive. Although the town's children "were not forgotten by Santa Claus," most of the striking miners had by this time been evicted from their company-owned housing. Many families had managed to find temporary homes and the union was building a large hall to accommodate those who had not yet found new places and to serve as a permanent home of the UMWA.[58]

The solidarity of the strikers, however, cannot be reduced to the influence of the union, important though it was. Two other factors must be taken into consideration. First, over the course of earlier disputes with the Dunsmuir empire in 1877, 1883, 1890-91, and 1903, miners had developed a high level of class consciousness. This found expression in the growing support for unionization and in the rise of working-class political parties. Without this, the UMWA would not have received the support it did. Second, the strikers were able to present a united front for almost two years because of the extraordinary depth of community support and the strength and commitment of the mining families. Large-scale strikes, such as the Island coal strike, are rarely simply confrontations between management and employees; rather they are best viewed as examples of collective action in which the entire community is mobilized.[59] This was recognized by union leaders, the local socialist MLAs, and the strikers themselves, all of whom spared no effort to encourage residents in the strike zones and to gather public opinion on their side.

Women were crucial to the mobilization of community support for the union cause. Confronted with the disintegration of the household economy and the increased domestic stress of trying to feed and clothe their families on meagre strike pay (strikers received $4 per week, their wives $2, and $1 per child), women took immediate steps to alleviate the psychological and economic burdens of unemployment and to ensure that striker morale remained high. This meant that they were often forced to assume additional roles in order to guarantee the survival of their families, such as finding jobs outside the home or becoming involved in community-based forms of family assistance, whether child minding or setting up soup kitchens. Women also sustained the morale of the community by organizing social events, including Saturday night festivals and weekly dances. According to one newspaper, women spared no effort to "plan and scheme innocent diversions to keep their menfolks in good humor."[60] This was not always an easy task. Most women would have been unaccustomed to having their husbands at home during the day, which must have added to the daily burden of household chores and child rearing. Many breadwinners, now idle, would gather in public spaces – street corners, benches, pubs – to talk and pass the time, although with the onset of fine weather in March and April, lumber camps and other forms of employment began to open up, absorbing some of the strikers.

Community solidarity was also enhanced by the strikers' belief that their cause was just. This appeared to be reinforced by the report of the commission of inquiry to investigate "generally into labour conditions of the Province,"[61] which the provincial government reluctantly agreed to launch on 4 December in the face of increasingly critical public opinion. Although the commission examined all industries, its primary focus, because of the strike, was the coal mines. Initially the striking miners were skeptical about the commission because it was reported that it would not even be stopping at Cumberland and Ladysmith.[62] Much to the surprise of the miners, however, the report was unusually sympathetic to their cause. Unlike the subsequent federal commission, the provincial inquiry examined many of the most important grievances of the miners, including the demand for union recognition, the docking system, wages, the cost of powder, and the need for washhouses at Extension and Nanaimo. Arguing that "no class of worker is ... more entitled to generous pay for a fair day's work than the coal-miner," the report supported unions, recommending that "no employer should be at liberty to discriminate against a workman because he belongs to a union."[63]

The proceedings of the investigation were a source of satisfaction for the UMWA and the striking miners. According to Farrington, who commented on the inquiry in the union's newspaper, the miners' representatives at the commission had successfully articulated the grievances of the striking men: "Frugal living, industrious and practical men who have spent all their working days in the pursuit of mining, told of cruel oppression, iniquitous robbery, infamous abuses, official arrogance and soul-racking drudgery encountered during their struggle to win a mean livelihood from this company. Poorly ventilated mines, dangerous working conditions, excessive cost of living, exorbitant rentals, short weights, monthly pays and lax, or non-enforcement, of mining laws, are only part of the cycle of evils surrounding these men." Just as significant, as far as Farrington was concerned, was that not only had this exposé of the reality of work in the Island mines been written into the official record but it had gone "unrefuted by the company."[64]

Perhaps the most important conclusion of the report, especially with regard to the status of the UMWA in Canada, was that there was no "definite evidence" to support the companies' claims that "the authority exercised by the union officials resident in the United States may produce conditions injurious to Canadian industrial interests."[65] This was a great fear among local industrialists, who believed that international American unions were coming to dominate labour in British Columbia and that the union was in collusion with producers in Washington state, who would benefit greatly from a strike in the Vancouver Island coalfields.[66]

Concern about the Washington state mines was a red herring, designed to alienate public support for the miners. Of much greater concern to union officials and the striking men was the fact that the Nanaimo-area collieries, especially Western Fuel at Nanaimo and the Pacific Coast colliery at South Wellington, had not joined the strike and were still working. In fact, they were benefiting from the strike because they were supplying CC(D)'s traditional markets, and therefore undermining the position of the strikers. Early in the strike, the union had recognized that its ability to pressure the provincial government and the coal companies into a settlement would improve if all the mines on the Island were shut down. As a result, by the spring of 1913, the union began to make plans for a strike at the Nanaimo collieries.

The expansion of the strike to Nanaimo on 1 May 1913 was a crucial aspect of union strategy, and marked a significant escalation in the conflict. May Day, the international day of labour, was a festive occasion in spite of the long strike. Past celebrations had been held in Nanaimo, but because of the unrest and the cost of travel, miners in Ladysmith and Cumberland decided to hold events in their own communities. The strike, of course, lent the celebrations a special significance, as they became a very poignant and public display of community solidarity. Special trains were chartered to bring Nanaimo families to Ladysmith, where a program of sports,

The Ladysmith May Day parade, 1913.
*Photo courtesy of Ray Knight*

games, and dancing was scheduled for children and adults. A May Queen, Ella Ball, was proclaimed and, "seated in a little dog cart, drawn by a tiny Shetland pony, and followed closely by two pretty girls, carrying a victory banner," she led a procession. The May Queen and her maids were followed by the children of the striking miners, carrying a banner that read "The Hope of the Future." Bringing up the end of the parade, which was reportedly attended by five thousand people, were the miners and their wives. They too carried banners and signs: "Workers of the World Unite" and "In Union Is Strength." When the parade reached the sports ground, the crowd gathered to listen to two hours of speeches by union leaders Robert Foster and George Pettigrew, local socialist MLAs Jack Place and Parker Williams, and Christian Sivertz, president of the BC Federation of Labour.[67]

The main event of the day, however, was not the parade and the speeches at Ladysmith but the mass meeting of about 1,500 Nanaimo district miners at the Princess Theatre later that evening, where union officials asked the miners to endorse the union's decision to shut down the Nanaimo mines. The decision to expand the strike to Nanaimo was enthusiastically supported by the majority of workers and by local political leaders, although it was later condemned by the Price Commission (Report of the Royal Commissioner on Coal Mining Disputes on Vancouver Island) as emanating "from an outside and, so far as this country is concerned, irresponsible authority who at least was not solicitous for the welfare of the industries of Vancouver Island."[68]

At the Princess Theatre, union officials, including Foster, Pettigrew, and Joe Angelo, argued that only through united action against all the companies could they compel the companies to enter into a collective agreement with the miners. Not all in the crowd agreed, and the meeting became quite tumultuous. At issue were two important points. First, collective agreements already existed at the Western Fuel Company in Nanaimo, at Pacific Coast Coal Colliery at South Wellington, and at the small Vancouver-Nanaimo Coal Mining Company at Jingle Pot.[69] Second, some miners insisted that the question of the strike be put to a vote of the employees of the three Nanaimo-area mines. The union officials argued that a strike vote was not necessary as about two thousand miners had voted to strike at the second UMWA convention six months earlier.[70] This did not satisfy all of the men, and on the following day another meeting of miners was called by a joint committee of Nanaimo, South Wellington, and Jingle Pot miners.

The joint committee meeting, led by Harry McKenzie, who had been shouted down the previous evening, condemned the union for not acting in accordance with "British fair play," and was determined that a vote take place. Owing to the hostile reception of the crowd, a vote resolution could not be passed. In the end, the committee, acting on its own authority and without the support of the majority, declared that a vote would take place at the courthouse the next day, 3 May.[71] Despite the fact that Western Fuel closed down to enable its men to vote, only 478 of approximately 2,150 district miners cast ballots. Of that number, 432 men voted to continue working. Harry McKenzie and the Price Commission later argued that the union had forced a strike upon an unwilling workforce, and that the reason for the low turnout at the courthouse was that the union had intimidated and frightened the men. The *Nanaimo Free Press,* however, accurately captured the mood of the miners when it wrote that the majority of men were "practically unanimous in its attitude of defiance to the necessity of a ballot."[72] Rather than being intimidated by the union, it is clear that the majority endorsed the union's decision to strike.

Given the low turnout at the polls, McKenzie, representing the joint committee, "agreed to refrain from work and advised those who voted to govern themselves accordingly."[73] They did. Within weeks, about 1,800 miners in the Nanaimo district had joined the UMWA.[74]

The decision to strike was endorsed by local MLAs Parker Williams and Jack Place. At a rally held at the Nanaimo Opera House on 11 May, they explained their position to the large crowd. While Place, as was his inclination, argued that the miners were engaged in a class struggle and stood to "gain by the overthrow of the capitalist system," Williams stressed the need to maintain unity and took specific aim at a number of accusations made against the striking miners and the union. He first addressed the issue of the existing agreement at Nanaimo, which, he argued, did not guarantee a living wage, clearly favoured management in hiring, and, because the agreements at the different mines expired on different dates, was designed purposely "to keep the miners disunited."[75] Furthermore, as the *BC Federationist* noted, the agreement severely restricted attempts by miners to resolve grievances with management. "No discussion of working terms or wages was ever admissible," and the attitude of the company was one of "take it or leave it."[76]

The controversial question of American interference in Canadian labour affairs had been raised a number of times over the past few weeks. On 9 May, the *Nanaimo Free Press* argued in favour of trade unionism, but rejected industrial and international unions. It consequently revamped the old idea of the formation of a Vancouver Island Miners' Federation. This was echoed a few days later by Harry McKenzie, who thought that if the miners united themselves in such a union, "they should be able satisfactorily to attend to their own business without the need for foreign interference."[77] This notion was endorsed by none other than the staunchly anti-union firm Canadian Collieries (Dunsmuir), which launched local 15 of the Dominion of Canada Miners' Union in late May 1913.

Vilifying the "foreign agitators" and attempting to play upon latent patriotism, CC(D) announced that it was willing to recognize a Canadian union, provided the miners renounce the UMWA. The labour press responded to this generous reversal of policy with abuse and ridicule. "So that the members may get the very best results from their union," the *UMWA Journal* announced, "the officers of the company have kindly consented to act as officers of the union. According to the last report the membership consisted of twelve faithful blacklegs," despite the fact that the initiation cost was only twenty-five cents and monthly dues ten cents.[78] Jack Place and Parker Williams also thought that this was an absurd proposition. How could a local union, with at most 4,000 or 5,000 members, muster the equivalent resources of the UMWA, which had a membership of almost 400,000? Like the MMLPA a decade earlier, it could only ever be a company union.

If the issue of a local union was a nonstarter for miners, the issue of foreign interference had a decisive impact on public opinion and had to be addressed. Williams discussed the issue of American involvement, pointing out that not only were capitalists permitted to combine but that the Western Fuel Company was an American-owned operation. But the issue went deeper than this because of persistent reports that union officials were in the pay of Washington state operators.[79] Part of the problem stemmed from an article written by Frank Farrington in March 1913, which sought to explain the causes and the "importance of this contest" to his predominately American audience. What motivated the union, he claimed, was the issue of competitive equality, a key policy of the executive. Essentially, competitive equality meant that in the interests of the unionized workforce, no coal-producing region should have an unfair advantage over another. The Vancouver Island companies, however, were in a much more advantageous position than the unionized mines of nearby Washington state. They were worked by non-union labour; they employed cheaper Asian labour; and the quality of coal produced was so high and demand for it so great that coal mines such as Washington's Roslyn and Cle Elum and the unionized mines of the East Kootenays could not compete, even though a forty cent per ton duty was charged on the importation of Island coal.[80] Hence, people argued that by bringing production to a halt on Vancouver Island, the union was eliminating the competitive advantage of the Island collieries, to the obvious benefit of Washington coal-mining operations.

As Parker Williams explained, however, the union was not trying to destroy the Island's coal-mining industry for the sake of Washington miners. Rather, it was attempting to raise the wages of Island miners to the same level as that of the unionized men in Washington state. Despite the fact that Washington miners produced an inferior product that sold for $2 less a ton, they received higher wages than Island miners. Indeed, they feared that if wages remained low on the Island, then Washington operators, arguing that they were unable to compete, would have a legitimate excuse to lower wages. That was why the UMWA was paying $6,000 per week in strike pay in Ladysmith.[81]

Despite these charges, the union remained firm and the men stayed committed to their cause. The overwhelming support of the Nanaimo miners boosted their resolve and changed the complexion of the dispute. The decision to strike at Nanaimo thus marked an important turning point. In many respects, it represented the high point of UMWA influence and organization on the Island. Able to present a coherent set of arguments that spoke to the many long-standing grievances of the miners, the UMWA was rewarded with a remarkable display of unity and solidarity. The UMWA's effectiveness in mobilizing the miners struck fear in both the government and the coal companies. Rather than attempt to solve the festering dispute, however, they became even more determined to fight the union. As a consequence, as spring turned to summer on Vancouver Island, conflict and unrest increased as the companies intensified their efforts to import strikebreakers and to stabilize production, and the government sent in more special constables to protect the owners' interests.

# 7
# *No Ordinary Riot*

The Great Strike is best remembered for the unrest that hit the strike-bound communities during the week of 12 August 1913. The violence, minor when compared with the deadly confrontations that occurred in mining towns in West Virginia and Colorado during this same period, was nonetheless unprecedented on the Island. The worst damage to property took place in Extension, South Wellington, and Ladysmith, where homes and a few company buildings were looted and burned to the ground. A number of people were physically assaulted, but there was no loss of life. One arrested miner, twenty-one-year-old Joseph Mairs Jr., died later of tuberculosis in prison, and one strikebreaker, Alex McKinnon, suffered serious injuries when dynamite was thrown into his Ladysmith home. If the scope of the unrest has been exaggerated in some accounts, so too was the response of the authorities. News of the unrest provoked an otherwise disinterested government into immediate action. After months of inactivity, it acted swiftly and harshly by mobilizing the militia and occupying the affected regions. Within a few days, almost a thousand soldiers had pitched tents in the strike zones, including a large contingent in Cumberland, which had not witnessed any serious disturbances. With the assistance of the provincial police force and the special constables, the militia drew up lists of suspected rioters and union organizers, who were promptly arrested. When the dust finally settled, authorities had arrested 213 men, tried 166, and sentenced 50 to prison terms. Of the 50 convicted men, 38 were from Ladysmith.

Contemporaries and modern historians have regarded the violent episodes of August 1913 in Ladysmith and Extension as the defining feature of the strike. In the minds of many, the unrest was nothing less than mob rule and anarchy, views that reinforced the common belief that miners were "scoundrels," "savage and vengeful brutes," only a few steps away from barbarians.[1] For recent commentators of all ideological persuasions, those hot August days also provide a provocative glimpse of the revolutionary potential of the Island miners. According to John Norris, for instance, "the climate of opinion in the labouring force was changing quickly from resentful acceptance of the economic helotry in which the companies held them to violent rebellion against the companies and the social order they represented."[2] This has been echoed by Allen Seager, a historian sympathetic to the miners' cause, who argues that the strike "brought a militant district to the brink of armed struggle on the eve of World War I."[3]

These interpretations, while flattering perhaps, are a serious misreading of the events. To argue that the miners were rising in violent rebellion against the existing social order completely ignores the objectives of the miners and grossly exaggerates the disturbances. Although some participants argued that they were waging a class war – MLA Jack Place, for instance, argued that victory in the strike would be the first step in a broader class struggle[4] – the strikes and riots were not inspired by a revolutionary political agenda. Instead, most miners were struggling for more mundane but equally important bread-and-butter issues, such as union recognition, improved wages, and better underground working conditions. If a class war was being waged, it was being waged by the government and companies. Perhaps Parker Williams, MLA for Newcastle, captured the sentiment of the majority of men best when he claimed that the strike "was not a fight between capital and labor but a fight against the minister of mines ... This is a fight not primarily for a living, but for the right to live. This is a fight against the undertaker and the morgue and the department of mines more than against the Canadian Colliery."[5]

During the early months of the strike, the miners and their families had largely restricted their activities to relatively peaceful mass street protests against strikebreakers. In Cumberland, the chief constable described the typical protest: "While conducting workers to their homes, the usual mob of about five hundred men, women and children followed, playing discordant music, shouting, yelling, and waving umbrellas, and otherwise behaving in a very disorderly and disgraceful manner."[6] Following the shutdown of the Nanaimo works, these activities intensified, and the frustrations of the striking miners mounted. The position of their families was deteriorating more and more as the strike continued and the economy went into recession, limiting other opportunities for employment. Some families left Vancouver Island altogether. In addition to increasing the number of Asians at work, the companies greatly intensified their efforts to import strikebreakers from BC and abroad, a task made easier by the growing number of unemployed. By June more than 300 men had been enticed to the Island from as far away as Durham, England, and Missouri. Despite tight security and deception, union officials often managed to inform these men that the strike was still on, and many refused to work. Those who did choose to work were often housed in the cottages of recently evicted strikers and were protected by a force of special constables, which had grown to almost 200 in number.[7]

Belated attempts by the federal government and the Vancouver Trades and Labour Council to negotiate an end to the strike were unsuccessful. The much-heralded visit to the Island in July 1913 by T.W. Crothers, the federal minister of labour, confirmed miners' suspicions that the government's intervention would not be effective. As one striker put it, it seemed that his "sole business here was to try and bunco the men back to work."[8] According to another report, Crothers spent remarkably little time with the striking miners. At Cumberland, he met five strikers at a local hotel, but "found time to interview seventeen of the Durham miners, brought out by the operators, and also some forty others working underground." It was the same at Extension, where he spent two hours with management in the company office but did not have a minute to spare the eighty-five striking men who had waited all day to meet with him. To make matters worse, throughout his tour of inspection, he refused to investigate claims that Canadian Collieries (Dunsmuir) was violating the Coal Mines Regulation Act, claiming that he did not believe the grievances of the men and that their "forceful protest" was wrong. In the end, he concluded that he had no hope "that anything I might be able to do will end the present lamentable situation."[9] His law partner, Samuel Price, was assigned the task of preparing a Royal Commission on the strike, and also visited the strike zone during July.[10]

If anything, the Crothers mission to the strike zone was counterproductive, as it further encouraged Canadian Collieries and the provincial government to take a hard line against the union. Their agents, the special constables, were becoming increasingly aggressive; just prior to a union meeting in Nanaimo, to cite only one example, a dozen mounted police rode straight into the gathering crowd of miners, forcing them to scatter.

Then on 19 July in Cumberland, during Crothers's trip, the worst outbreak of violence to date occurred. For about a week, there had been rumours that strikebreakers, led by a man named Cave, were going to drive the strikers out of town. On the evening of 19 July, CC(D)'s payday, a procession of strikebreakers marched into town determined to "clear out the strikers," at which point a crowd of striking miners began to gather. According to one account, "Cave, along with about twenty more, paraded up and down the street, with the arrogance of a rooster, sure of no opposition, throwing out challenges to the strikers, who would not accept the temptations and provocation until he had insulted a lady."[11] A general melee ensued, at which point a group of about fifteen special constables on horseback charged into the crowd, urged on by company cashier G.W. Clinton, who, according to Robert Foster, "ordered the special police to ride us down, shoot us, we were not good enough to live."[12] Following the charge, Cave continued to antagonize the strikers and when the miner Jack Muir took up his challenge, Muir was arrested. A week later, six of the strikers, including the popular local union president, Joe Naylor, who had sought to keep the strikers from confronting Cave, were arrested. The chief of police at first refused to investigate Cave, who had instigated the disturbance. Only when Alexander Campbell, the new mayor of Cumberland, issued a warrant for his arrest was Cave taken into custody. Following numerous adjournments, the case against him was finally heard several months later on 5 September; much to the disappointment of the assembled crowd, the charges were thrown out by the local magistrate.[13]

The temperature in the strike zone was clearly rising. Four days after the July incident, there were reports that someone had attempted to blow up the Trent River bridge, located six miles from Cumberland, causing minor damage. The police suspected that the perpetrator came from a group of strikers employed on government road work near the bridge, and immediately requested reinforcements.[14] In an atmosphere of rumour and speculation, however, not everyone believed the police. Cumberland's mayor even went so far as to deny publicly that any such attempt had been made to destroy the bridge. "On three different occasions," he wrote, "it has been reported that the Trent River bridge, over which it is necessary for the Company's shipments of coal to go in order to reach Union Bay, was blown up ... Has any serious damage been done to this structure so far? The miners here treat these reports simply as a joke."[15]

These events had a dramatic impact on the mood of the striking miners. Although the miners engaged in picketing and "rough music," the strike had for the most part remained peaceful and nonviolent, as Mayor Campbell explained in an open letter to the press following the August riots: "The fact should be continually borne in mind that at no time since the beginning of the strike have the Police Commissioners or the City Council seen any occasion in which it was necessary in maintaining the peace, or preventing the destruction of private property." Given the "destitution, provocation [and] insults from within and without" that accompanied the "invasion" of the special police in September 1912, he continued, and the "lack of assistance or sympathy from those who have the power to cause an investigation into their grievances," it was "a remarkable fact that these men have so much control over themselves as to respect the law."[16]

The presence of the special police aroused anger, frustration, and fear in the miners, largely because they were armed and wore no badge or any other identifying mark; they had quickly become the odious symbol of arbitrary state authority. According to the Pinkerton agent, reporting to the attorney general in late August 1913, the strikers' "hatred for the specials is something indescribable."[17] Aggressive, arrogant, poorly trained, and, according to some, otherwise unemployable, the specials were deliberately provoking the miners through their abusive language and violent behaviour. There were numerous incidents of police harassment, intimidation, and violence. Following the mounted charge in Cumberland, a striker by the name of Mandzik was arrested for vagrancy. While in jail, he was beaten so badly by the officer that he was unable to stand unsupported during his trial. Although the original charge was dismissed, Mandzik was subsequently charged with assaulting his jailer.[18] In his reports from Nanaimo, the agent pointedly emphasized that "the specials are at fault in many instances," because they continually antagonized the strikers and their families. These men "should be compelled to maintain silence," and "must avoid retort and maintain the same dignity that the military observe." Ultimately, the agent felt obliged to telephone the chief of police, who "at once decided to take steps to check the practice among specials."[19] Their behaviour contrasted somewhat with that of the regular police, who, as members of the community, often knew the strikers personally and were more sympathetic to their cause.[20]

As Cumberland simmered, the situation in Ladysmith and Extension also deteriorated. By July more rumours of impending trouble were beginning to circulate. From Stark's Crossing, a short distance from Ladysmith and Extension, John Stewart wrote to the attorney general that he had heard men saying that "there will be murder and bloodshed in that camp if the strike lasts much longer,"[21] and following the 20 July arrival from Vancouver of a transfer barge with strikebreakers on board, local police noted a significant increase in the number of confrontations between the strikers and the "working men." These incidents were still relatively minor, ranging from the use of "bad language while in bathing" to preventing CC(D) employees from unloading freight, although there were rumours of more serious incidents, for example, that strikers were tampering with railway switches.[22] There was also a marked increase in street protests and demonstrations against the strikebreakers. In Ladysmith, as in Cumberland and Nanaimo, men, women, and children daily accosted strikebreakers and their police escorts at the train station, resulting in numerous arrests and charges. Threats were also apparently made against Ladysmith's doctor and pharmacist because they continued to treat strikebreakers.[23] In Nanaimo, two women were fined $20 for calling the pit bosses "scabs." The women, the magistrate warned, surely risked losing "the respect of their husbands" with such behaviour.[24]

Rather than take positive steps to ameliorate the explosive situation, the company and the state took actions that seemed purposely designed to provoke the miners. Some of the strikebreakers, escorted by the special police to and from work, were provided with firearms and other weapons. As in Cumberland, this made the strikebreakers in Ladysmith more confident and aggressive. According to the Pinkerton spy, they were now "beginning to resent any slurring remarks thrown at them" and did not hesitate to fight back.[25] This lack of restraint was revealed on 5 August, when police in South Wellington arrested two "foreign" strikebreakers for threatening miners' pickets with a shotgun. They were later fined $50 and costs.[26] Four days later, another incident occurred at Ladysmith: four strikebreakers, reported to be Southern Italians, attacked two of the strikers with a knife, stabbing James Hatfield. The police arrested the knife-wielding strikebreaker but refused to arrest the other men involved. This incensed the strikers. At a meeting held in Ladysmith to discuss the incident, the men decided that the time had come to take a stand: "that since they had in the past done their best to preserve peace and had been successful, unless they were to be protected by the police they had no alternative but to protect themselves." Only at this point did the local police reluctantly agree to arrest the men.[27]

Against this background of continual police harassment and growing frustration at the miners' weakening position, the tension that had been slowly building up in the strike zones during the spring and early summer exploded during the second week of August. The unrest began in Nanaimo on Monday, 11 August, and spread the next day to South Wellington before culminating in the much more serious riots in Ladysmith and Extension on 13 August. The decisive factor, once again, was the presence of strikebreakers protected by the hated special constables. Upon hearing rumours that a large contingent of strikebreakers was heading to work at Nanaimo's No. 1 mine, between 500 and 700 strikers and their families began picketing the streets and the pithead. At one point, the crowd collected outside the home of a local strikebreaker named Pattison, who along with other members of his family had returned to work that day. Whether out of fear or as a warning, Pattison "poked a gun through the wire screen of a window in a threatening and menacing manner."[28] The police managed to subdue and disperse the crowd before there was any serious incident, but this set the stage for future confrontations.

The following day, 12 August, crowds of miners and their families once again picketed No. 1 and subjected the workers to a liberal dose of rough music. Local MP Frank Shepherd tried to appease the crowd but offered little of substance to strikers apart from declaring his continued support for trade unions. Manager Thomas R. Stockett also appeared, claiming that the men were underground only because a section of the south wall of No.1 was on fire; no coal was being mined. However, he had no response to the crowd's suggestion that he apply to the union for the necessary men.[29] Afterwards, there were a number of scuffles between the strikers and the special constables, in which car windows were smashed and a few strikebreakers, including Tully Boyce, one-time president of the Miners' and Mine Labourers' Protective Association (MMLPA) and current opponent of the United Mine Workers of America (UMWA), suffered minor injuries.[30]

Whether they were inspired by events in Nanaimo or acting on their own initiative, striking miners at South Wellington, where the Pacific Coast Coal Mines Limited was attempting to increase production, also decided that the time had come to take matters into their own hands. Reinforced by miners from Nanaimo, a group estimated at between 600 and 800 people, marched through the town. The small contingent of special police was unable to control the crowd, which, according to press reports, "broke up property and beat up the men now working at the mines, these being subjected to great violence, as a result of which it was deemed advisable to rush their wives and families into the bush for safety."[31] The crowd then descended upon the carpenter's shop, where the strikebreakers were being housed, and also attacked the Chinese bunkhouses, the homes of strikebreakers, and a few other company buildings.[32] Having driven the strikebreakers from the town, the crowd subsided and an uneasy calm prevailed.

While company officials searched the woods for the strikebreakers and their families and the company physician set the broken bones, news of the unrest in South Wellington reached Ladysmith; by about 11:00 that night, the situation became tense as a crowd of people, estimated at between 300 and 400, mingled outside the Temperance Hotel, where about fifteen strikebreakers were living.[33] At midnight, a boat arrived carrying five new strikebreakers and the paymaster for Canadian Collieries. As these men slowly made their way to the hotel from the harbour, they were met by a procession of angry demonstrators, some of whom had been drinking. One miner, Charles Axelson, was arrested by the police for singing, "Hurray, hurray, we'll drive the scabs away." But the real trouble began when Axelson's wife was told of her husband's arrest. This "stout lady" – a "veritable Amazon in build, vigour and strength" – proceeded at once to mount a rescue operation. Wielding an axe, she forced her way through a line of "awestruck" police and special constables and freed her husband from Ladysmith's jail.[34]

Mrs. Axelson and her newly liberated husband were met by a large crowd of supporters, which then proceeded once more to the Temperance Hotel, where jeering and singing strikers began to hurl rocks at the windows of the building. At about 12:30 A.M., a small explosion occurred when a stick of dynamite was thrown at the back of the hotel. The demonstrators then marched up and down Ladysmith's streets. The homes of some electricians and engineers who worked for CC(D), as well as those of some strikebreakers, were stoned.[35] An hour or so later, another explosion was heard along First Avenue, at the home of strikebreaker Alex McKinnon. A bundle of dynamite had been thrown through the window of a bedroom where five of his children were sleeping. McKinnon, hearing the sound of glass shattering, apparently rushed into the bedroom and grabbed the makeshift bomb. Unable to throw it through the broken bedroom window because of the blinds, he ran with it into the kitchen. Just as he was trying to hurl the dynamite back outside, it exploded, blowing off his right hand and causing serious damage to his face and body.[36]

This attack, the most violent act to occur during the strike, "caused consternation and horror in the community where the crime is denounced as dastardly."[37] People immediately assumed that it had been the work of striking miners, and a Pinkerton agent was recruited from Seattle by the attorney general to investigate the crime and to learn of the union's future plans. The agent was unable to unearth the culprit, however. The miners, he reported, informed him that they thought it was the work of "foreigners," or that McKinnon had blown himself up as he was trying to throw dynamite at the demonstrators outside his home. The spy did not believe them and strongly suspected that they knew who had committed the crime but were hiding that information from him.[38]

In the end, when arrests were finally made in August 1914, a full year after the event, people were surprised to learn that the crime had been committed by three men, none of whom were miners, who had been "passing a bottle round" that evening: Bill Stackhouse, variously described as a barber or "poolroom owner"; "Tango" or "Tangle" Jackson, bartender at the Grand Hotel; and a stranger named Mike Adams. After the explosion at the Temperance Hotel, they had apparently decided to attack McKinnon's place in order to "shake them up." McKinnon, who had drawn strike pay for almost a year, had recently repudiated the union and returned to work. He was also accused by some of deliberately provoking the strikers: "When coming home from work at the mines he would open his lunch bucket and offer the crusts of his lunch to the picketing strikers, bread that had been bought with money he had received from the union before he went to scab. Naturally this did not sit well with the strikers although they did not take him seriously enough to do anything about it."[39] Although Stackhouse had been arrested and prosecuted for his part in the riots – he was apparently "in this trouble from the beginning to the end. He was in the mob at the Temperance Hotel; he marshalled the men who marched through the town, calling one section 'G. Company,' drilling and giving orders and acting as a leader. He was in the riot at the wharf when the office was attacked, and was at Lapsansky's"[40] – he had never been a suspect in the bombing. Only when he turned himself over to the police in August 1914 after his release from prison was the case solved.

As dawn broke on 13 August, about 400 strikers and their families began to parade through the streets of Ladysmith. Later in the day, Ladysmith's mayor informed the *Victoria Daily Times* in a telephone interview that "the town is now controlled by the mob and it is impossible for the six men who constitute our police force to cope with the situation ... The mob is patrolling up and down the streets, attacking non-union workers and smashing windows in their houses." He had therefore urgently requested the assistance of the military.[41] Newspaper reports indicate, however, that it was "quiet all day,"[42] although the miners were no less determined to drive away the strikebreakers. According to authorities, they allegedly formed a "peace mission"; instead of using force, they used threats to persuade the strikebreakers to stop working by going around to their homes and telling "each one that if he remained after [12:00, 15 August] he would not be safe, and that if he wished safety in the interval he must go to the Union Hall and sign a paper."[43]

Union leaders Sam Guthrie and Joseph Taylor tried to keep the demonstrations peaceful and were asked by the police and local civic leaders to persuade the strikers to go home. Clearly worried about an escalation of the violence, they also met with merchants and businessmen, imploring them to "use their endeavours to keep out of the town the militia and special police"[44] in order to avoid more protests. The only serious incident that day occurred when strikers raided Ladysmith's Chinatown, where they vandalized property and "ordered all the Orientals out of the place." According to the *Vancouver Sun,* "John obeyed complacently, packed his goods and made for Victoria."[45]

As tempers cooled in Ladysmith during the day of 13 August, unrest flared up elsewhere in the strike zone. In Nanaimo, the situation reached a climax around noon when the miners met about nineteen strikebreakers at the train station. The strikebreakers, tired and battered, had been driven out of South Wellington the night before. The crowd of strikers marched them down to the CPR docks, where they were to be shipped out of town on the *Princess Patricia,* which was scheduled to dock at noon. They reached the dock just as the ship arrived. On board was a group of twenty-five special constables. As they descended the gangway, they were confronted by a crowd of angry miners threatening to "dump the Sons of

Bitches into the bay." Having forced the specials back on board the ship, the miners confronted a provincial policeman who had been observing the situation unfold. In the confrontation, Jack Place took away the constable's revolver, which he had been wielding menacingly, in an attempt to defuse the situation and prevent possible bloodshed. Overwhelmed, the police were put on board the *Princess Patricia* and unceremoniously sent back to Vancouver.

As the crowd of miners returned to town, they were informed that there had been a major confrontation at Extension between strikers and strikebreakers in which six strikers had been killed. Although it was later learned that this was not true, about a hundred men armed themselves and proceeded to march en masse to Extension.[46] The previous day there had been a standoff at Extension as strikers, reinforced by men from South Wellington, prepared to confront the strikebreakers. In response to the picketing strikers, the manager had armed many of those working at Extension, who then braced themselves in the pithead compound known as the bullpen for the inevitable confrontation. Throughout the night, the situation had been volatile. Alcohol flowed freely in the bullpen, and strikebreakers antagonized the strikers by shining a searchlight on the homes of union men. The following afternoon, the strikers, their numbers now augmented by men from Ladysmith and South Wellington, gathered once again at Extension. After being refused a meeting with the manager, they advanced on the bullpen around 4:00 in the afternoon, at which time the armed strikebreakers opened fire.

It was at this point that a messenger was sent to Nanaimo with the erroneous report that six men had been killed in the exchange of gunfire. Among the Nanaimo miners who set off for Extension were Constable Hannay and UMWA officials.[47] When they arrived on the scene, they too were shot at by the strikebreakers, but were too numerous for the men in the bullpen, who either retreated into the tunnel or fled the scene entirely. Amazingly, only one person, a contractor from Nanaimo by the name of James Baxter, who had gone to Extension "to see the fun," was shot when he approached the mouth of the tunnel.[48] When dusk settled on the area, a group of miners turned their attention to the village. Their targets apparently picked out ahead of time, the strikers ransacked and burned homes and buildings belonging to the company, the strikebreakers, and the Chinese, driving their inhabitants into the woods.

Coal cars destroyed during Extension riots.
*BC Archives, A-03162*

When dawn broke on 14 August, much of Extension lay in ruins. Although only one man had been injured in the fighting, there was considerable property damage.[49] Some of the strikebreakers were still in the tunnel of the mine, but many had managed to flee the area with their wives and children. About 1,500 men from Nanaimo, South Wellington, and Ladysmith were now in the town; although there was some sporadic looting, an eerie calm prevailed. That afternoon, a contingent of provincial militia arrived on the scene from Departure Bay, where they had landed earlier in the day. Armed with two Maxim guns and ten thousand rounds of ammunition, the soldiers clearly feared that they were entering a combat zone. They had expected to come across a bloodbath and to find that "men were buried alive." Much to the surprise of Lieutenant Colonel A.J. Hall, this was not at all the case. "The whole business has been grossly exaggerated," he claimed in an interview with the *Vancouver Sun*. "All the wild stories of shooting and loss of life are without a shadow of truth. There has been a little rough work and perhaps some horseplay but there has been really no danger of life." The newspaper continued, stating that Hall's "report to government headquarters would be that trouble was not nearly as bad as had been feared, and there was comparative order for a district where a strike was raging."[50]

Militia pose in the remains of manager J.H. Cunningham's house at Extension.
*BC Archives, D-03287*

Once at Extension, the troops extinguished a fire at the pithead and mounted a "rescue" operation, rounding up about a dozen strikebreakers and their families who had taken refuge in the woods, and searched for some strikebreakers who were still hiding in the tunnel. Under the protection of the police, they were then taken to Victoria, where they were maintained at public expense. Having rendered aid to the civil power, Hall's troops then returned to Nanaimo, where peace and order also prevailed. The following morning, 15 August, the crowd at Extension took advantage of the temporary absence of the militia to burn down the house of J.H. Cunningham, the manager of Extension, who had fled to Port Alberni the moment the violence began. That same day, more troops arrived in the strike zones from Vancouver and Victoria. Although there had been no disturbances in Cumberland, 450 soldiers were sent there to keep the peace, and a further 130 were stationed in Ladysmith, bringing the eventual total to 1,000.

The arrival of the militia in the strike zone has been celebrated by local historians, who regard the military's aid to the civilian power as a glorious chapter in the province's military history. Several myths predominate. First, popular writers such as David Ricardo Williams maintain that the militia deterred further violence in those regions threatened by the breakdown of civil order and anarchy. This is based on the assumption that the miners were committed to a long-term course of violence, and that this was indeed the beginning of a rebellion or uprising that threatened the existing social and political order and that needed to be put down by military force. However, not only was the violence over by the time the militia arrived on the scene but there is no evidence to suggest that it was going to continue. The second myth is that the presence of the militia was welcomed by the general public. Once more, according to Williams, "the business community of Nanaimo wholly supported their presence, and even amongst the miners there was a certain grudging respect."[51] On the contrary, from the outset, the arrival of the militia was highly controversial. The striking miners and the union certainly did not view the militia as a "positive presence." Rather than showing "grudging respect," strikers and their families jeered and ridiculed the arrival of the "tin" soldiers and the "beardless boys." Jack Kavanagh, vice president of the BC Federation of Labour, probably expressed the feelings of many when he claimed at a mass meeting of union members in Vancouver that Nanaimo was being "terrorized by half-clad barbarians armed with rifles and bayonets ... No reptile ever evolved from the slime of ages resembles the spawn of filth now on Vancouver Island and known as the militia."[52]

Middle-class observers also criticized the extreme, oppressive tactics of William Bowser, attorney general and acting premier in the absence of Richard McBride, who on 12 August had left on a journey to London. In a statement made after the order to send in the army, he declared: "It is the business of the militia to preserve order, and we are going to do it in Nanaimo and the other places if we have to call out every militiaman in the country."[53] Dubbing him "Napoleon Bowser" for behaving as though

"Napoleon's Retreat from the Bonaparte," *Vancouver Sun,* 15 August 1913.

a state of war existed on the Island, the Liberal opposition press denounced his heavy-handedness, declaring that "the government's action in ordering out a thousand men to do the work of a hundred has caused much ill feeling among the business men of the city" of Nanaimo.[54] Many merchants and civic leaders also found the very presence of so many troops undesirable. In Nanaimo the soldiers were criticized for "loll[ing] about in the sunshine," for trying to censor the press, and for excessive drumming and parades that interfered with local traffic. Moreover, concerns were voiced about the cost of maintaining the army's presence. Tradesmen worried about who was going to pay them for the supplies the militiamen were requisitioning, and the press reported that it was costing $12,000 per day to keep the army in the field.[55]

The presence of the military was also largely ineffective, and by no means the stunning success it is often claimed to be. Rather than destroying the union and compelling the miners back to work, Bowser's heavy-handed approach – incredibly, the soldiers were equipped not only with bayonets but also with Maxim guns and even an "armoured train" consisting of a rail car carrying a large field gun – provoked more ridicule and abuse than fear from the population. And arguably, rather than restoring order, the militia actually helped the miners' cause. Not only did the military occupation arouse public sympathy for the strikers but in the short term, at least, the miners prevented the strikebreakers from working, driving them from the mines. Others would eventually take their place and go to work, but only under the heavily armed protection of the remaining soldiers.

Indeed, the entire affair was like a poor theatre performance, with the miners and their supporters watching the various parades and processions, especially those of the "half-clad barbarians," with a mixture of bewilderment and amusement. The commander of the troops, Lieutenant Colonel Hall, condemned the local press for revealing the "strategic moves" of

The "half-clad barbarians" of the 72nd Regiment pose with their Maxim gun.
*BC Archives, A-03195*

the militia, and suggested that reporters should "be provided with military uniforms [and] be attached to a corps and have all their news censored by an officer."[56] Soon the lieutenant colonel's concern about keeping his troop movements secret was overshadowed by a more serious problem. The militiamen, many of whom were clerks and shopkeepers, were reluctant to participate in Bowser's campaign in the first place, and were becoming anxious to return to their civilian lives. There was little for them to do, they were not being paid the wage of $2.50 per day promised them by Bowser, and many feared losing their regular jobs. By the end of August, there were threats of desertion.[57]

Those soldiers not lolling about in the sunshine were kept busy enough. Soon after the arrival of the militia, the authorities began to draw up a list of suspects and warrants for their arrest. Some voiced the fear that the arrests might lead to further rioting, but this was unfounded and the militia and police soon began rounding up suspected rioters. The first arrests were made in Nanaimo on 18 August, when the militia surrounded the Athletic Hall, where a crowd of about 1,200 miners was meeting to discuss the recent settlement between the union and management at the Jingle Pot mine. With the Maxim gun pointed at the back door of the hall

Soldiers of No. 1 Company, 58th Regiment "lolling" about in the sunshine in a water tank at Extension, August 1913.
*BC Archives, D-03283*

and bayonets drawn, the militia ordered the men to exit the building in groups of five or six; those singled out as ringleaders and troublemakers were promptly taken into custody. After the arrests, the militia tore the hall apart searching for nonexistent weapons.[58]

The excessive force of the militia and the mass arrests at the Athletic Hall, which included a number of Jingle Pot miners who had just voted to return to work, sparked protests from provincial labour leaders and UMWA officials but did not stop the work of the police.[59] The dragnet continued. Chris Pattinson, Joe Angelo, and Joseph Taylor, all union organizers, were soon taken into custody. On 8 September, George Pettigrew was also arrested. By 19 August, 128 men had been arrested and were sitting in the Nanaimo jail. That same day, 48 men appeared before Nanaimo magistrate Simpson under a heavy guard of militiamen with fixed bayonets. Simpson set the tone of the hearings and the subsequent trials when he refused to grant bail to the men, who were charged with crimes ranging from unlawful assembly, assault, obstruction of the police, and picketing. Among those arrested were two boys, one of whom was only fourteen years old, and local MLA Jack Place. They were remanded in custody for eight days prior to the start of preliminary hearings. Place's arrest caused something of a scandal. He was charged because he was in possession of the revolver of constable Harry Taylor, which he had taken from him during the disturbances at the Nanaimo dock in order to prevent the violence from escalating further. Worse, however, was that he was subjected to a "long tirade" against him by Magistrate Simpson, for being in dereliction of duty: "You are to be censured more than any person taken off the street," Simpson declared. "Your action is not that of a decent law-abiding citizen."[60]

By the end of August, 71 men from Ladysmith and Extension had been arrested and charged with rioting and other offences, although some were released and not all were committed to trial. As with the Nanaimo men, those arrested in Ladysmith were held on remand and denied bail until the trial. There were a number of serious charges. Sixteen-year-old William Bowater Jr. was charged with burning down Cunningham's house at Extension. Of greater concern was the charge of attempted murder laid against another teenager, Edward Morris, who was accused of trying to kill J. John, a strikebreaker at Extension.[61] As in Nanaimo, many of the arrests were clearly arbitrary and politically motivated. No strikebreakers were ever arrested for their part in the gunplay at Extension, yet the

Arrested strikers being marched through Ladysmith.
*BC Archives, E-01194*

police rounded up local union officials and tried to silence and humiliate local politicians. Unlike Jack Place, Parker Williams managed to avoid arrest, although he was rumoured to be on the "waiting list," because he had been looking after his children while his wife was in hospital at Ladysmith. His son did not have such a watertight alibi. David Williams was arrested and charged with unlawful assembly.

The trials of the Ladysmith men began in Nanaimo on 9 October and were presided over by New Westminster judge Frederick W. Howay, who was brought in to replace the local magistrate. The Ladysmith miners were subjected to the full measure of the law. On the advice of their lawyers, they chose trial by judge and pleaded guilty to the charges, believing that they would receive lighter sentences. They were soon disabused of this idea. Arguing that the men had committed acts of "terrorism," Howay handed down harsh sentences. He divided the defendants into three groups – ringleaders, the "general run of the rioters," and those only "slightly interwoven" in the unrest. The "ringleaders," including Joseph Taylor and Sam Guthrie, president and secretary of the union local, respectively, were sentenced to two years' imprisonment, although neither man had been involved in the rioting.[62] Three other youths, William Simpson Jr., Paul Deconinck, and John Morgan, were also identified by Howay as ringleaders and received two years' imprisonment. The twenty-three "general run" rioters were sentenced to one year and fined $100 each; the remaining eleven men were sentenced to three months' imprisonment and a fine of $50 each.

The trials of the Ladysmith men were controversial. The harsh sentences were condemned in the media, as was Howay's obvious hostility towards the miners. Consequently, the trials of the Nanaimo and Extension men were held in New Westminster, where they were presided over by Judge Aulay Morrison. The "Strikers' Assizes" lasted from November 1913 until late March 1914 and contrasted greatly with the "Speedy Trials" held in Nanaimo. After spending months in prison awaiting trial, most of the men were either acquitted or freed from jail, having already served time. In March, most were given amnesty by Premier McBride – in response, at least in part, to pressure from miners' wives and families, who began petitioning the premier in January. In April, twenty-two Ladysmith men – minus the young Joseph Mairs, who had died of tuberculosis in prison – were released from Oakalla penitentiary under the same conditions.[63] When the prisoners returned to the Island, they were feted by large crowds singing the *Marseillaise.*[64]

In the aftermath of the riots and arrests, the miners in Nanaimo and Ladysmith settled into an uneasy routine. Although the mood was sombre, the majority of the strikers remained defiant and the union determined. In Nanaimo the men congregated at various locations, such as the post office square or in taverns, to discuss developments among themselves. They also attended meetings and sought part-time work to sustain their families and reduce the tedium of unemployment. Apart from harassing the odd strikebreaker, most of the men refrained from doing anything to provoke the militia or the special constables who kept them under constant watch.[65] When the Western Fuel Company decided to reopen No. 1 mine with non-union men on 25 August, the union picketed the entrance, giving rise to speculation about further violent confrontations.

However, the attempt to restart the mine fell through, as "no men came out to work, and all local miners are proving loyal to the union."[66] The determination of the strikers was also evident in the parades and processions that accompanied the transportation of the arrested miners. When a large number of men were transferred to Victoria from the overcrowded jail in Nanaimo, "at every station along the line they were met by large crowds and cheered enthusiastically as martyrs and heroes of the cause."[67]

While the August unrest was a key event in the labour dispute, its impact on the course of the strike was actually quite limited. The riots provided the authorities with an unparalleled opportunity to end the strike, but Bowser's invasion of the strike zone failed to crush the union or destroy the miners' solidarity. If anything, the riots and the military occupation reinforced the strikers because these events mobilized the entire community. The presence of heavily armed soldiers, the arrest of local union officials, and the mass trials disheartened the striking miners and their families, but they quickly recovered their composure and accommodated themselves to the new situation.

The Pinkerton spy correctly reported that the strikers were behaving like "chastised children" and posed no threat of further violence – after all, they naturally feared further arrests. However, rather than the back of the strike being broken, as Wargo claimed,[68] the miners were not silenced or forced back to work. On the contrary, buoyed by the recent agreement between management and workers at Jingle Pot and by continuous rumours of an expected settlement at Nanaimo's Western Fuel Company, they continued to protest and confront strikebreakers – albeit not so aggressively as during the hot summer months – confident that they would eventually emerge victorious in the dispute. Even in late November, after the sentencing of the Ladysmith men, union support remained as strong as ever. As the dejected government spy reported from Nanaimo: "It is almost certain that not over two dozen men have quitted the ranks of the union to resume work since the critical period which followed the passing of sentence here by Judge Howay and which so nearly approached breaking the strike ... The men are far from giving up the battle."[69]

From the outset, people speculated about the cause of the August unrest. Most government and company officials, as well as middle-class commentators, for whom the violence had come as a great shock, automatically assumed that the disturbances had been planned in advance by the union. This was most clearly expressed by Frederick William Howay, the presiding judge at the Ladysmith trials, who maintained that the events were no "ordinary riot ... one of those sudden ebullitions of pent-up feeling, but it shows all down the line a course of deliberate scheming and planning; there is a design in it from one end to the other."[70] Almost without exception, recent historians have endorsed the judge's conspiracy theory, arguing that the riots were premeditated and organized by the UMWA. For example, Lynne Bowen, who states in her discussion of Howay's judgment that "no such premeditation had been proven or even mentioned during the trial," nonetheless concludes on the basis of the presence of union officials at the demonstrations and riots that "just as in Nanaimo and South Wellington, the Ladysmith riot was probably organized by the union beforehand as a desperate measure to force a settlement."[71] John Norris also maintains that the rioting was planned by lower-level union officers "as a means of resolving their impasse."[72]

Some contemporaries also saw more sinister forces at work, blaming the riots on foreign agitators. The provincial police suggested that "the situation was much aggravated by the fact that the ranks of the strikers were augmented by an invasion of sympathizers from outside points; and this suggests that the outbreak was organized in advance."[73] This was a powerful argument, one that salved wounded pride and guilty consciences, and enabled people to indulge in latent anti-Americanism. Perhaps the companies were right. After all, the UMWA was an international, American union that some felt did not have the interests of the Canadian worker at heart. From the beginning, its actions were considered "un-Canadian" and offensive to the strong sense of "British fair play." Had not the local press been pointing out for months that the only ones to benefit from the strike were Washington state coal miners who had not come out in support of their Island brothers as promised? Had not Frank Farrington (the highest-ranking member of the union's international board) himself argued about the need for "competitive equality?" Norris echoes this sentiment when he claims that "the implementation of [the rioting] was undertaken by a group of intermediate-level organizers, brought in by the union from the United States to provide picket captains and muscle where this was needed ... Canadian strike leaders apparently had only a minor part in the organizing."[74]

Especially worrisome to the authorities were the largely unsubstantiated rumours, repeated by the press, that members of the radical working-class organization, the Industrial Workers of the World (IWW, or "Wobblies"), had incited the violence. Apparently, scores of these foreign troublemakers had been able to "pass the local immigration men with little trouble." Even if there was no proof that they were involved in the rioting, one newspaper maintained, they should still be deported as "undesirables."[75] Although Norris rightly argues that the "IWW was a convenient bogeyman rather than a serious direct influence on the strike," he nevertheless suggests that the movement's indirect influence "may have been more considerable."[76]

Evidence of IWW involvement in the strike and the rioting, whether direct or indirect, is murky at best. Robert Gosden, prominent IWW spokesman and future labour spy, apparently "strode boldly into the fray," advocating industrial sabotage and other forms of direct action.[77] Were Wobblies responsible for the attempt to blow up the Trent River bridge? There were also reports that two Wobblies had been caught trying to set fire to the wharves at Union Bay, but according to the mayor of Cumberland, one was now in the employ of Canadian Collieries and one had been told to leave the area.[78] Did the Wobblies incite the riots? Isaac Portrey, the miner who had been on the gas committee at Extension with Oscar Mottishaw, appears to have either had some connections with the IWW or developed them later. He was arrested for participating in the demonstrations, and later lived and worked with Gosden.[79] In November 1913, Gosden achieved a measure of respectability when he was elected president of the Miners' Liberation League, on the basis of a speech in which he advised McBride and Bowser, much to the pleasure of the crowd, "to employ some sucker to taste their coffee in the morning before drinking it if they value their lives."[80]

For the most part, however, the IWW appears to have been inactive during the strike. In Nanaimo on 15 August, a few Wobblies "were understood to have made an attempt to organize a mob from among themselves to go down to the wharves and ridicule the militia," but according to the newspaper, this effort fell flat.[81] In the aftermath of the riots and Gosden's speech, rumours continued to abound of Wobblies arriving from Vancouver "for the sole purpose of dynamiting and shooting," but nothing seems to have materialized of this either.[82] Although the Pinkerton agent reported in October that the strikers approved of the arrival of two IWW agitators, who were going to work in Nanaimo No. 1 "with the intention of creating trouble there among the men already employed,"[83] the UMWA and the majority of the miners never embraced the tactics or agenda of the IWW. Shortly after the riots, Farrington criticized press reports suggesting that the IWW was involved. They were designed "to connect socialists and IWWs with the miners' strike for the obvious purpose of attaching to the miners any odium that may be felt towards these two movements, and to detract from the merits of the miners' strike." The UMWA, he claimed, was "a non-political organization, composed of men of all shades of political opinions, and has no relation whatever with the IWWs."[84] By the end of the year, the agent was reporting that there was "no more mention of the IWW help which some time back was said to be forthcoming to take up the work of destruction."[85]

Other reports denied that the unrest was fomented by foreigners. One policeman, who had been an eyewitness to the Extension riots and had been so badly beaten that he was reportedly "close to death," argued emphatically in perhaps the most perceptive analysis of the events, that he had been "maltreated" by "miners purely and simply. They were not men imported for the occasion, as we have been told since we came to Vancouver. They were English and Scotch and Irish miners, exasperated beyond endurance and acting apparently from a sense of their own personal wrongs."[86] In fact, there is very little evidence that the unrest was fomented by foreigners or that it was organized in advance by the union. Farrington and other union executives were in Vancouver, where they were negotiating a settlement, mediated by ex-MLA James Hawthornthwaite, with the Vancouver-Nanaimo Coal Mining Company at Jingle Pot. Did they orchestrate the riots in the belief that they could then force Canadian Collieries and Western Fuel to recognize the union and accept a collective agreement, as historians have argued? This seems highly unlikely, as the violence could easily have scuttled the talks. Indeed, from their perspective, the rioting undermined the union's position and harmed its public image.

The presence of local union officials in the crowd, particularly Joe Angelo, an Italian American union organizer at Extension, and Louis Nuenthal at South Wellington, also does not prove that the union organized the violence. The union often hired ethnic organizers like Angelo, and they often acted independently. Furthermore, the official UMWA policy was one of strict nonviolence.[87] It knew that violence rarely if ever benefited the strikers, but instead played into the hands of the companies. Thus, when the union struck at the Nanaimo collieries in May 1913, it demanded that the men "exert every effort to prevent unlawful or abusive tactics ... during this contest,"[88] knowing that violence only invited retaliation and repression, not to mention legal costs, and often resulted in the loss of the strike.[89]

Reaction to the unrest from labour groups and the union reflected this general disapproval of violence. The labour press argued that the miners had been incited to violence by the actions of the company and the state, but nonetheless called the destruction at Extension the result of "poor judgement." Instead, it suggested that "a much better showing could have been made against the mining company by destroying the coal tipple and thus crippling the mine. This apparently was not touched, although a number of hovels in the vicinity were destroyed."[90] Frank Farrington was more categorical in his condemnation of those miners who had participated in the August disturbances, declaring at a meeting of the Vancouver Trades and Labour Council: "I don't regard the men who are now under arrest for rioting at Nanaimo as martyrs at all. I regard them as fools who had not sense enough to keep their mouths shut. Now some of these fellows think it is fine to get in the limelight. They are anxious to get their names in the paper and be cheered as heroes, but as a matter of fact a man behind bars is not doing very much to help the cause."[91]

Accusations of foreign incitement and union organization of the riots, not to mention those of mob rule and anarchy in the strike zones, also exclude one important factor when explaining the unrest: the striking miners themselves. In all these interpretations, the actual men and women who participated in the demonstrations and rioting are ignored, as are their grievances, frustration, anxiety, and anger. Instead of being active participants, they are unjustly reduced to the status of passive pawns of nefarious manipulators, a view reminiscent of that of our Pinkerton spy, who compared the men to "a band of sheep and who display just about the same initiative as would those animals."[92] Norris essentially repeats this argument when he claims that "the rank and file followed along without much question."[93] Hence the *non sequitur,* voiced by the spy, that the arrest of the small number of troublemakers brought the disturbances to an end.[94]

Rather than being an episode of mob rule or an example of organized rebellion against the existing order, the riots are best characterized as a form of social protest designed to restore the perceived moral balance of society, the economy, and the community. Such forms of violent social protest are not at all unusual in modern industrial society, although they are often regarded as irrational, anomalous events, the relics of a preindustrial past that have since been replaced by strike action, considered by many to be a less violent and more rational and modern form of protest. Rather than being a relic of traditional society, however, violent social protest has been an essential component of modern industrial society; equally important, it has often been successful. Furthermore, collective and individual violence, coercion, and intimidation were not just features of working-class protest. On the contrary, they were just as likely, if not more so, to be employed by companies and the forces of the state, the police and the military, in order to prevent or end labour disputes.[95]

Riots and crowd actions are complex events that cannot simply be reduced to "spasmodic" responses to external stimuli or be treated as examples of mobile anarchy, as they were portrayed in the local press. Instead, they must be understood as highly rational and disciplined forms of collective behaviour that occur within the context of deepening social distress over a period of time. Unlike mob rule, such forms of social protest are governed by a clear set of objectives that contain within themselves a "legitimizing notion," which validates the action of the crowd as a form of self-defence, and require the consensus of the community.[96] In fact, the very act of participating in a crowd action, in which individual identities are subsumed in the common cause, reinforced within the community a sense of wholeness and solidarity, however momentary and effervescent.[97]

In the case of the most serious riots in Ladysmith and Extension, but also in the minor scuffles that occurred in Nanaimo, the primary objective was to "drive the scabs away." The animosity of the crowd was directed towards three specific groups of strikebreakers: those who had been imported from outside; those who had initially supported the strike and received strike benefits, but had since chosen to return to work; and the Chinese. Special constables were often caught in the middle but were not specifically targeted. In Ladysmith, strikebreakers' homes were pelted with stones and some men were physically assaulted. This was the case at Extension as well, although the crowd also targeted company property. The main difference at Extension was the presence of a large force of strikebreakers, armed and huddled together in the compound or bullpen at the pithead, determined to fight if the strikers advanced on them. The primary issue was not that the strikebreakers were allowing the operators to resume production, although this was certainly a great concern. Rather, it was the conviction that the strikebreakers – protected by the state while the strikers had literally been abandoned by it – posed a direct threat to their livelihoods and to their future and that of their community. This was especially the case for those strikers, like Alex McKinnon, who had returned to work after having drawn strike pay for many months. In choosing to work, they were undermining the collective rights of the miners – the right to a fair wage, the right to organize, the right to safe working conditions – and were literally taking food out of their families' mouths. They were like shirkers in wartime. Rather than performing their collective duty, rather than making a sacrifice for the common good, they were exploiting a crisis for their own personal benefit. As such, they outraged the moral sense of the community.

The collective action was not a spasmodic act but a deliberate response to months of provocation, frustration, and growing anger that built upon a general consensus within the mining community. This consensus set limits on the actions of the strikers; in effect, they were licensed by popular opinion to drive the scabs away, but nothing else. Hence, the bombing of McKinnon's house caused general outrage within the community. (It is worth repeating that the men who were eventually convicted of the offence were not miners.) The popular consensus was created by shared convictions and experiences during a time of meaningful, legitimate struggle, and was based on the belief that traditional social values, rights, and identity were being threatened. In other words, a sense of communal self-interest had coalesced among enough people at once, to validate the actions of the crowd, create a unified, cohesive front, and, momentarily at least, overwhelm any potential source of opposition within the community.

The wives and families of the striking miners were instrumental in creating this consensus. Not content to watch the strike from the sidelines, the wives, daughters, and sisters of the striking miners, many of whom were members of the United Mine Workers Women's Auxiliary, organized social events, held informal political education discussions, organized weekly strategy meetings, planned protests and demonstrations, raised much-needed funds for the union and the families of the strikers, and took to the streets.[98] The scope of women's involvement was revealed during the trials of the Ladysmith men. "At Ladysmith and at Extension," a surprised *Vancouver Sun* reporter announced, "a great deal of the damage done was the work of women sympathizers."[99] They threw bricks and stones through the windows of strikebreakers' homes, called out abuse, and assaulted those strikebreakers they happened to come across. The reputation of the strikers' wives during the rioting at Extension was such that many of the wives and children of the strikebreakers fled out of fear into the woods.[100] However, not all of them did. Charlotte Schivardi, the wife of a strikebreaker at Extension, caused a sensation when she admitted giving a strikebreaker a gun. She "played as militant a part as a leader of her people in the strike troubles as she has in these trials," reported the *Sun*. Whereas union women were invariably condemned as "amazons," Schivardi was admired for her bravery by the media and heralded as a "non-union Joan of Arc."[101]

Although women participated in "riotous assemblies," Judge Howay was unwilling to punish them directly. In order to set an example to the women and the rest of the community, he purposely handed down harsher sentences to the men. He admitted as much when he indicated that he might have been more lenient in sentencing the men had it not been for the actions of their wives. When counsel for the miners requested mercy on the grounds that their wives and children would suffer, Howay replied that "they know that that is a plea that would strike a sympathetic chord. But what do I find in the evidence? Woman is sympathetic and kind, that is one of the features in her character that most appeals to us, but what do I find? I find in this case that the women, and in many instances your wives, were there in the crowd, singing 'Drive the scabs away,' throwing stones themselves and urging on the work of destruction. That takes very much from the strength of any such plea."[102]

No women were arrested for participating in the riots, but they interpreted the actions of the Crown as a declaration of war not just against the unions and the working class but also against women and their families.[103] In contrast to the men, who were generally subdued in the wake of the riots and the military occupation, the women continued to agitate and confronted the specials and the soldiers, the odious symbols of state authority and repression, with renewed vigour. The Pinkerton reports reveal women's hostility towards the police and the militiamen, especially during the trials and the sentencing of the miners. They were "the most outspoken and bitterest of enemies to the soldiers" and were "often cautioned by their men folks when speaking in reckless tones of the soldiers."[104] In September, when a train carrying arrested miners arrived in Nanaimo from Victoria, where they had been held without bail, their wives and daughters attempted to pass packages of food, tobacco, and clean clothes to them, but were prevented from doing so by the soldiers. When they eventually reached the courthouse, the procession of women started to vilify the militia, and stopped only when the chief of police threatened to arrest them. When the miners returned to the train station later that same day, the women were still "out in force and waylaid the escort and kept up a running fire of talk and bother until the depot was reached and here as the men were entrained occurred a wild ten minutes during which the soldiers were driven to extremes to preserve a line of formation and hold back the crying, cursing and hystirical [sic] amazons who almost fought for a chance to incite trouble."[105]

Scenes like this were repeated in Ladysmith. When the Pinkerton spy travelled there to take in the preliminary hearings, he expressed indignation at the Ladysmith miners, who were clearly "a distinctly lower order of people and display the traits employed by the ignorant and savage. Foreigners seem to predominate." But once again he especially noted the attitude of the women, whose actions so brazenly violated every conceivable norm of "civilized," "feminine" behaviour. "One can readily see that no act is too mean or contemptible for them to stoop to." At one point, they attacked a female witness for the Crown in a "sickening display of savagery," and later, when another witness described the property damage at Extension, the women in the courtroom were "particularly exuberant" and "chattered and laughed like Amazons" while "their men folks seated in the rear tried to quiet them."[106] Clearly, neither the militia nor the trials intimidated the women. Unable to visit their husbands and sons who were in prison awaiting trial, the women petitioned, and vociferously protested to, the attorney general for visiting rights. This request was refused on the grounds that "plots for further disturbance may be concerted during those visits." The authorities, mused a writer for the *Vancouver Sun,* "have cause to fear the women."[107]

The fear expressed by the authorities appears to have been due less to the potential violence of the women, which could easily have been contained, than from the threat posed to established gender relations and the patriarchal status quo by the women's behaviour. The protests and petitions were an intensification of a broader process of politicization of working-class women on Vancouver Island that was being increasingly expressed in the demand for civic equality and the right to vote.[108] The call for the vote during the Great Strike reflected a groundswell of opinion in working-class communities and organizations. Universal suffrage was now endorsed by the socialist parties, the Vancouver Trades and Labour Council, and, in early 1912, by the BC Federation of Labour.[109] It was also supported by some middle-class reformers, although mainstream middle-class society trivialized the demand for equal suffrage. A reporter for the *Vancouver Sun* was baffled at women's apparently sudden and new interest in "such an abstract question as votes for women." This was evidently the result, he thought, of the overabundance of "leisure" time in the striking miners' households.[110]

The demand for the vote was neither an abstract question for women nor simply an unforeseen consequence of idle chitchat, but reflected the politicization of women in the mining communities. They linked political reform to the broader questions of social justice and class conflict raised by the strike – issues that went to the heart of the community and the family. Of particular importance was mine safety. The nightmare of the Extension explosion remained especially vivid. "Up there," indicated one woman in early 1913, "is the burying ground watching and waiting, saying ... who's coming next. And you never know who it's going to be. Maybe it's my boy she is waiting for. She got my husband."[111] One woman, "whose enemies call her the suffragette," claimed, "Who wouldn't fight if there were laws made for the safety of the miner and they were not enforced, how'd you like to go to the bowels of the earth if you wasn't sure at what minute an explosion might come?"[112]

Similar sentiments were expressed by an unnamed miner's wife and mother of ten children, who declared in February 1913: "Do you think if the women of the province had the vote and elected representatives to the house that they would allow those men not to care whether the law was not enforced and our menfolks had to work where they were risking their lives every minute? No they wouldn't. Women are different. They wouldn't ask how much will it cost to make the mine safe. They would say those men who toil way below the earth's surface must be protected."[113] Mine safety was not just a question of working conditions and workplace control but a question of the continued health and well-being of the family and community. "You don't forget when you see 30 graves all new dug

"Miners' Lives of Secondary Consideration," *Vancouver Sun,* 7 February 1913.

in a row waiting to be filled with men you've known all your life," one woman remarked. "Wouldn't you fight, and starve if need be, if when your man left the house you didn't know how he was coming back?"[114]

If the demand for the vote struck many as revolutionary, the women of Ladysmith and Vancouver Island were hardly following a radical feminist agenda. Although they demanded political equality, like middle-class suffragists, they accepted separate spheres for men and women. Despite the "militancy" of the women and the fact that they were carving out new roles for themselves in the community, their goals were essentially conservative and defensive. They sought to protect the family, not to challenge men's dominant role in the economy and in politics.[115] The vote would naturally expand women's role in the public sphere, but for the miners' wives it was a means of defending, not escaping, the "domestic sphere."

Despite the fact that women were fighting to protect, not destroy, their place within the domestic sphere, the strike undermined traditional notions of domesticity and challenged gender roles. Many women were forced for the first time to seek employment outside the home and to assume new burdens of responsibility within the household economy. Men had little choice but to accept this new reality. Moreover, women's recourse to direct action threatened not just the patriarchal norms of middle-class society but the security of male dominance within the working-class family. As men's position as sole breadwinner was eroded, so their "manhood" was diminished; as women forged new relations and roles for themselves, they redefined the space of authority within the family and the community. This was unacceptable to the authorities and representatives of the middle-class public, which refused to prosecute them, and to many of the women's husbands, who tried to silence them.

The arrival of the militia was seen as a declaration of war against the striking miners and politicized the dispute, as the state, rather than just the mining companies, became the focus of working-class agitation throughout the province. This was embodied in the creation on 6 November 1913 of the Miners' Liberation League. The league consisted of representatives of the most prominent working-class organizations in British Columbia, including the Vancouver Trades and Labour Council, the British Columbia Federation of Labour, the UMWA, the Industrial Workers of the World, the Socialist Party of Canada, and the Social Democratic Party of Canada. The primary purpose of the league was "to secure the liberation of the miners who had been sent to jail for their connection with the labor troubles in Nanaimo and district," and in the coming months it organized rallies and petitions, and raised much-needed funds through an effective pin campaign. It considered running labour candidates for the provincial legislature, but gained its greatest political notoriety through the influence of the IWW and the election of prominent Wobbly Robert Gosden as the league's first president, who advocated direct action and general strikes.[116]

Gosden's fiery rhetoric and calls for violence were at great odds with the UMWA and the majority of miners, which despite the August debacle, continued to pursue a negotiated settlement to the dispute. However, the unrest had undermined the union's efforts and discredited it even more in the eyes of the companies. When the UMWA decided to apply to the federal department of labour for an arbitration board under the aus-

BC Miners' Liberation League Tag Day, 20 December 1913.
*Photo courtesy of Ray Knight*

pices of the Industrial Disputes Investigation Act in November 1913, the companies refused to participate. They justified this by pointing out that the act was inoperable after a strike or lockout had been declared, but clearly the companies had no desire to negotiate a settlement that, among other things, would have entailed recognition of the union. The companies, the *Nanaimo Free Press* observed, "point out that the strike has now been in progress for eighteen months, that the mines are in operation and that the men are no longer in the employ of the companies."[117]

Even pressure from Premier McBride in the spring of 1914 could not move the companies to negotiate an end to the strike. The companies' position seemed to be unassailable. By employing Asian labour – the number of Chinese workers increased from 315 in 1912 to 432 by May 1913 – and strikebreakers from Alberta, the United States, and England, Canadian Collieries and Western Fuel had enough workers to maintain adequate levels of production in a depressed market.[118] As Bowen writes:

"Mine owners, summoned before McBride on March 17th [1914], declared themselves quite happy with things as they were. Production figures were climbing ... Besides, the coal market was depressed, Cumberland mines were on short time; the last thing the mine companies needed was more employees. Let the strike go on, the owners would resist the UMWA forever."[119]

Despite the controversial and much-publicized arrival in early June 1914 of a famous American union activist, the octogenarian Mother Jones, the strains of the two-year battle were beginning to tell on the miners and the union. In a series of speeches to overflowing crowds in the striking communities, Mother Jones condemned the tactics of the companies and the state and advocated a general strike.[120] Buoyed perhaps by her bold rhetoric, on 23 June the miners rejected an offer on the old terms made by the companies under the pressure of the provincial government. Soon afterwards, however, rumours began to circulate that the union was going to cut off strike pay.[121] Less than one month later, the miners' fears came true when UMWA district president Robert Foster announced that the union would no longer pay strike relief.[122] The writing was on the wall. Mounting tensions in Europe meant that plans for a general strike, agreed upon in principle by the BC Federation of Labour convention on 15 July, were doomed to fail. On the evening of 19 August, just over two weeks after the First World War began, a majority of striking miners voted to end the strike. At the bottom of the front page of the *Nanaimo Free Press,* overshadowed by news of the war, a small announcement marked the companies' triumph and the end of an era.[123]

# *Conclusion*

The Great Strike was a watershed in the history of Ladysmith and the coal-mining industry of Vancouver Island. When coal was king, Ladysmith had thrived. As one writer put it, "the three T-wharves built to ship the coal stood as a symbol of never-ending prosperity."[1] This optimism and faith in the future of the town, however, were seriously challenged by the strike, the war, the decline of the industry, and the Great Depression. The miners and their families had built a community and an identity based on the shared experience of work, disaster, and struggle against the operators. When they overcame built-in divisions and competing interests, their solidarity could be impressive, but as the strike revealed, neither they nor their union were a monolithic force. In their struggles, they marched and protested, and they raised their voices in a call for economic and social justice, but, in a context hostile to reform, few listened. In the wake of the strike, defeated, demoralized, and divided, they fell silent.

The end of the strike and the beginning of the First World War brought little immediate relief to the strikers and their long-suffering families. The economic recession continued, and rather than increasing production, the onset of war caused it to decline initially. The depressed market for coal and the ready availability of strikebreakers meant that unemployment in the coalfields remained exceptionally high, a problem that was compounded by the blacklisting of many union supporters. No doubt the poor economy and hostile labour environment, and not simply patriotism, influenced the decision of many young men in the area to join the army. At least forty-one men from the Ladysmith district lost their lives during the Great War, outstripping the casualties of the Extension explosion of 1909.[2]

The recession in the coalfields did not last, however. The insatiable demand for munitions for the killing fields of northern France created a great need for coal. Blacklisted miners were eventually rehired and new mines were developed. But the war only temporarily masked the decline of the coal-mining industry. It had created an artificial boom; production at Extension, for instance, increased dramatically from a low of 57,855 tons in the strike year of 1913 to 256,952 tons in 1916, but never again reached the prestrike record of 331,576 tons in 1911. By 1926, production at Extension had sunk to 175,811 tons, while employment – 881 men in 1911 – had dropped to only 540 men. Many former Extension miners found work at the newly opened Morden, Reserve, and Granby mines, but the future of these operations was also by no means secure.[3] Restructuring of the industry – in 1927 Canadian Collieries (Dunsmuir) bought out the Western Fuel Company – and the steady increase in the use of fuel oil threatened the long-term viability of coal mining on Vancouver Island.[4] By late 1928, rumours were circulating that the mines were about to close, and on 25 March 1929, the Western Fuel Company announced that it was suspending work in Nos. 1, 2, and 3 mines at Extension. Two years later, in April 1931, as the world sank deep into economic depression, the Extension mines closed forever.[5]

The closure of the mines and the global depression had a devastating impact on Ladysmith. Many small businesses and industrial enterprises, such as the smelter, a stove works, a cigar factory, the two breweries, and the bottling plant, closed. The once busy industrial waterfront stood silent, and the population collapsed from a high of 3,295 in 1911 to about 1,400 in 1934, as people left in search of employment.[6] A number of schemes were developed to attract investment and work, but they all foundered. The small logging camps and the lumber mill in Chemainus could absorb only a few men. As a result, Ladysmith entered a period of grave financial crisis. The town's debt rose even as its tax base collapsed. Many residents were unable to pay their property taxes and many lots reverted to the city, which at the height of the Great Depression claimed title to more than three-fifths of the incorporated land.[7] Only with the arrival of the Comox Logging and Railway Company in late 1936 would Ladysmith begin the long, slow process of economic recovery.

The miners never recovered from their defeat in the strike. During the war, the position of British Columbia workers did improve somewhat as full employment gave them new opportunities to flex their economic muscle. Historians Allen Seager and David Roth even speak of the gathering "momentum of industrial conflict" that "continued into the postwar period and affected an increasing variety of workers."[8] Indeed, the number of strikes increased significantly after 1916, especially in the manufacturing and lumber sectors, and membership in the United Mine Workers of America (UMWA) grew impressively in the Kootenays and Alberta. The same cannot be said for the Island miners, however. If labour in general became increasingly militant in British Columbia and western Canada during the war years, leading to the eventual formation of the One Big Union and to the Western Labour Revolt of 1919, unionization and labour protest remained dead letters for Island miners.

Did the union fail the men, as Lynne Bowen has suggested?[9] It is difficult to agree with this conclusion, given the union's significant financial commitment to the two-year Great Strike. Besides, what else could the union possibly have done? Or did the men let the union down? John Norris concludes that the miners "had not yet moved to the more sophisticated level, desired by the international, of supporting patient organizational industrial unions to enforce the sharing of industrial decisions with management."[10] He then suggests that divisions within the workforce and the ability of the companies to resist the strike made the work of the union more difficult. Given the duration of the dispute, however, the miners' patience and their solidarity were its most striking features. Furthermore, one could argue that the union's more "sophisticated" and "patient" organizing efforts ultimately hindered rather than helped the miners' cause because they encouraged faith in compromise and settlement. Negotiated settlements require willing partners, however, and these the miners simply did not have. In an economic and political climate hostile to unionization, the combined power of the state and big business proved too much for the striking miners and the union. The real failure was society's refusal to accept organization as the workers' natural economic and social right.

Had the miners struggled in vain? The strike has generally been viewed negatively in light of its failure and the violent episodes of August 1913. Indeed, the events of that August have cast a long, dark shadow over the dispute, distorting the goals and accomplishments of the striking miners and their families. Violence in the Island coalfields was an exceedingly rare occurrence, however. The August unrest was the result of a particular set of circumstances that existed in the strike zone: the growing number of strikebreakers protected by the state; the unwelcome presence of special constables, who deliberately provoked the miners and their families; the hostile and confrontational government and companies; and the progressive impoverishment of the community. These factors produced a sense of extreme crisis, increasing the miners' frustration and leading inevitably to a tense, explosive situation. Rather than being the result of a conspiracy or an act of rebellion, the riots were indeed a spontaneous protest against existing conditions. There were of course political overtones, but they did not reflect the desire to overturn the status quo. Rather, their trajectory was more conservative, more a defence of the miners' traditional rights and customs than an attempt to foment revolution. Socialism became the political voice of the miners not simply because they were inherently radical or militant but because in the polarized political context of British Columbia, the parties of the left were the only political organizations that responded positively to working-class interests. More significantly, perhaps, the ultimate political objective of the unrest was internal, not external, designed more to mobilize and reinforce community solidarity than to send a message to the state. As such it had enormous implications for the internal dynamics of community relationships.

The notion of community is essential to understanding the history of small towns like Ladysmith and reveals the weaknesses of the theory of Western Exceptionalism, which focuses on external structural forces at the expense of the men and women who made up the communities. The foundations of community in Ladysmith were in place well before the Great Strike and the August unrest. Despite the transitory nature of earlier mining camps, such as Wellington, a strong sense of community developed in the new town of Ladysmith, spurred no doubt by incorporation and the permanence this suggested. Apart from the common experience of work in the mines, the most important factor contributing to the forging of community bonds and values was the long tradition of self-help, self-reliance, and independence that characterized coal-mining centres around the world. The proliferation of lodges, associations, and friendly societies ensured not just the economic viability of working-class families but their very survival. Over time, interaction reinforced a strong communal ethos; the shared values gave coherence to the experiences of residents and a common identity began to emerge that transcended social, economic, and ethnic status. This in turn gave miners a stake in their communities, a reason to struggle for their economic and political rights.

However, community is much more than the existence of a common, cohesive set of values and attitudes. Rather than being static, community is, as a number of writers have pointed out, a dynamic, historical process based upon continuous dialogue and interaction. Community, in other words, can perhaps best be described as the arena in which ideas and values are debated in an ongoing process of negotiation. This negotiation is seldom benign or nonconfrontational, because it operates within the parameters of existing power relations. Consequently, it is not always equal or devoid of conflict, but requires – indeed, demands – reciprocity and mediation if community is not to disintegrate into its constituent components. What makes community, therefore, is not the existence of consensus or a sense of common purpose in and of itself. Rather, community is the very process of negotiation and dialogue, the continuous working and striving towards a common purpose, the continual effort, as it were, to ensure that everyone is on the same page. As one historian has argued: "What is common in community is not shared values or common understanding so much as the fact that members of a community are engaged in the same argument ... the same discourse, in which alternative strategies, misunderstandings, conflicting goals and values are threshed out."[11]

Within the context of this arena or forum, the dialogue between the competing and complementary forces of ethnicity, class, religion, and gender unfolded in Ladysmith, producing multiple loyalties among the town's residents, shaping their identities in different ways, while at the same time forging communal life. The most obvious illustration of this is the role of women. Women's participation in the dispute reinforces the fact that strikes were not just between striking men, their unions, and the company officials, but involved the entire community. Without this support, the strike would not have lasted two months, never mind two years.[12] Through their mobilization and involvement in the strike and the riots, women were empowered both in the public and private spheres. If the implications of this were unsettling to the men and society at large, the emancipatory impact on the shaping of women's identity and political culture was nonetheless decisive, giving the shared experience of miners' families coherence while at the same time radically transforming and expanding the discursive framework of the community and society at large.

Within this context, the poignancy of the August unrest becomes clear. If community is the stage on which conflicts of gender, race, and class are played out, the violence was a turning point in its historical development, transforming, redefining, and renegotiating community in an attempt to overcome the immediate crisis and prevent its disintegration. Despite the ultimate collapse of the strike, it revealed an awesome potential, one that was not lost on the provincial authorities and elites.

# *Notes*

**Introduction**

1 Richard Goodacre, *Dunsmuir's Dream* (Victoria: Porcépic Books, 1991), 39.

2 *Census of Canada,* 1911.

3 There has been remarkably little written about this important strike. See Lynne Bowen, *Boss Whistle* (Lantzville, BC: Oolichan Books, 1982); John Norris, "The Vancouver Island Coal Miners, 1912-1914: A Study of an Organizational Strike," *BC Studies* 45 (1980); Alan J. Wargo, "The Great Coal Strike: The Vancouver Island Coal Miners' Strike, 1912-14," BA thesis, University of British Columbia, 1962.

4 Jeremy Mouat, "The Genesis of Western Exceptionalism: British Columbia's Hard-Rock Miners, 1895-1903," *Canadian Historical Review* 71 (1990): 317.

5 David J. Bercuson, "Labour Radicalism and the Western Industrial Frontier, 1897-1919," *Canadian Historical Review* 58 (1976): 154-75. See also A. Ross McCormack, *Reformers, Rebels, and Revolutionaries: The Western Canadian Radical Movement* (Toronto: University of Toronto Press, 1977); Paul Phillips, *No Power Greater: A Century of Labour in British Columbia* (Vancouver: Boag Foundation, 1967); and Martin Robin, *The Rush for Spoils: The Company Province, 1871-1933* (Toronto: McClelland and Stewart, 1972).

6 The frontier thesis has many difficulties, most notably the very concept of the frontier. It is often employed as a geographic term meaning a region that has not been settled or colonized, or as an economic term meaning a region whose natural resources have not yet been exploited. Underlying both definitions is the sociocultural assumption that a frontier is a region untouched by "civilization," those legal, cultural, social, and ethical codes of a colonizing, metropolitan culture. Not only does this ignore the presence of indigenous populations but the frontier thesis represents a legitimizing discourse of colonialism, expansion, and domination. A frontier is a frontier only from the perspective, and in the language, of a conqueror.

7 Bercuson, "Labour Radicalism," 154.

8 Ibid. These arguments also reflected an unsubtle political agenda as they clearly underwrote western Canadians' views about the distinctiveness of their society within Confederation. What is interesting about Bercuson's interpretation is not just the wholesale adoption of the frontier thesis to western Canadian historiography but that in reviving it, he implicitly and uncritically accepted its ideological baggage. According to one authority, Turner's famous thesis did not just celebrate "American uniqueness and the ideology of individualism"; it also had important practical consequences for historians concerned with their own status within an emergent profession still dominated by men and institutions from the American northeast. The frontier thesis "was generally appealing to American historians, but especially to those not on the East Coast," because it enabled the mostly Midwestern historians "to do regionally and locally celebratory work under the aegis of this paradigm, and let them do professionally

valued work with locally available archives ... Turner, among others, believed that the West represented distinct values ... There were, he wrote to a fellow mid-westerner, 'characteristic Western ideals and social traits' resulting from 'the experience of the frontier and of the democratic aspirations of the pioneer period'": Peter Novick, *That Noble Dream: The "Objectivity Question" and the American Historical Profession* (Cambridge: Cambridge University Press, 1988), 94, n. 14.

9 Clark Kerr and Abraham Siegal, "The Interindustry Propensity to Strike: An International Comparison," in *Industrial Conflict,* ed. Arthur Kornhauser, Robert Dubin, and Arthur M. Ross (New York: McGraw-Hill, 1954), 191. Isolation theory has come under a great deal of criticism. See, for instance, David Gilbert, *Class, Community, and Collective Action* (Oxford: Oxford University Press, 1992).

10 Brian Lewis, *Coal Mining in the Eighteenth and Nineteenth Centuries* (London: Longman, 1971), 66.

11 Bercuson, "Labour Radicalism," 168.

12 Mark Leier, "W[h]ither Labour History: Regionalism, Class, and the Writing of BC History," *BC Studies* 111 (1996): 66, comments, "Who made revolutions? Fools, according to Bercuson," in reference to Bercuson's book, *Fools and Wise Men: The Rise and Fall of the One Big Union* (Toronto: McGraw-Hill, 1978). See also Gilbert, *Class, Community, and Collective Action.*

13 See, for instance, Mouat, "The Genesis of Western Exceptionalism"; R.A.J. McDonald, "Working Class Vancouver, 1886-1914: Urbanism and Class in British Columbia," *BC Studies* 69-70 (1986): 33-69; James Conley, "Frontier Labourers, Crafts in Crisis, and the Western Labour Revolt: The Case of Vancouver, 1900-1919," *Labour/Le Travail* 23 (1989): 9-37; Mark Leier, "Ethnicity, Urbanism, and the Labour Aristocracy: Rethinking Vancouver Trade Unionism, 1889-1909," *Canadian Historical Review* 74 (1993): 510-34; Leier, "W[h]ither Labour History"; Mark Leier, *Where the Fraser River Flows: The Industrial Workers of the World in British Columbia* (Vancouver: New Star Books, 1990); Mark Leier, *Red Flags and Red Tape: The Making of a Labour Bureaucracy* (Toronto: University of Toronto Press, 1995). See also James Naylor, *The New Democracy: Challenging the Social Order in Industrial Ontario, 1914-25* (Toronto: University of Toronto Press, 1991).

14 John Douglas Belshaw, "Cradle to Grave: An Examination of Demographic Behaviour on Two British Columbian Frontiers," *Journal of the Canadian Historical Association* 5 (1994): 51, n. 38. Just how isolated the mining communities of Vancouver Island were in the late nineteenth and early twentieth centuries is open to question. Certainly geographic isolation from local urban centres existed, but this was relative, and how this contributed to radicalism and militancy is unclear. Cultural isolation is a different question altogether. Despite the distance from their metropolitan culture in Britain, British colliers arrived on Vancouver Island not only with their cultural baggage intact but probably with a heightened, exaggerated sense of their Britishness as the spearhead of Empire and the colonial project. Likewise, it seems illogical or at least inconsistent to support the concept of isolation while rejecting that of the frontier. Is not a frontier a frontier precisely because of its isolation from "civilization"?

15 Belshaw, "Cradle to Grave."

16 John Belshaw, "The British Collier in British Columbia: Another Archetype Reconsidered," *Labour/Le Travail* 34 (1994): 11-36.

17 John Belshaw, "The Standard of Living of British Miners on Vancouver Island, 1848-1900," *BC Studies* 84 (1989-90): 37-64.

18 M.J. Haynes, "Strikes," in *The Working Class of England, 1875-1914,* ed. John Benson (London: Croom Helm, 1985), 95.

19 *Report on Strikes and Lockouts in Canada from 1901 to 1912* (Ottawa: Ministry of Labour, 1913), 5-6.

20 Ian McKay, "Strikes in the Maritimes, 1901-1914," *Acadiensis* 13 (1983): 10, 15. McKay points out that official data on strikes in Canada are defective and make regional comparisons difficult. He derives his information from a variety of sources.

21 Ian McKay, "The Realm of Uncertainty: The Experience of Work in the Cumberland Coal Mines, 1873-1927," *Acadiensis* 16 (1986): 8, 52.

22 Ibid., 7-8.

23 Ibid., 20; Stephen Hickey, *Workers in Imperial Germany: The Miners of the Ruhr* (Oxford: Clarendon Press, 1985), 168, also notes this, writing that "miners have often shown themselves a historically aware profession, conscious of and knowledgeable about their links with earlier generations who toiled below the ground and who shared many of the same experiences. A sense of history has itself contributed to their collective development and to their struggles." David Alan Corbin, in his study of the miners of West Virginia, similarly captures the outlook of the coal miners: "The coal miner was not alienated from his work or product. He took pride in his career – once a miner, always a miner. He possessed a 'proud sense of occupational identity' that ... helped him to define himself and gave him an identity that seemed to be lacking among other industrial workers. He understood his materials, the work he did, the strategy of extracting coal, 'how his job fit into the overall pattern'. His daily travel underground gave him an overview of the entire operation, past and present. In an industrial nation dependent upon the energy resource he produced, the coal miner knew he was essential to the nation and that he was necessary for the common good": Corbin, *Life, Work, and Rebellion in the Coal Fields: The Southern West Virginia Miners, 1880-1922* (Chicago: University of Illinois Press, 1981), 39-40.

24 Norris, "The Vancouver Island Coal Miners," 68.

**Chapter 1: A Selfish Millionaire**

1 See, for instance, Terry Reksten, *The Dunsmuir Saga* (Vancouver: Douglas and McIntyre, 1991); Lynne Bowen, *Three Dollar Dreams* (Lantzville, BC: Oolichan Books, 1987); Eric Newsome, *The Coal Coast: The History of Coal Mining in BC, 1835-1900* (Victoria: Orca Books, 1989); James Audain, *From Coalmine to Castle* (New York: Pageant, 1955). See also D.T. Gallacher, "Men, Money, Machines: Studies Comparing Colliery Operations and Factors of Production in British Columbia's Coal Industry to 1891," PhD thesis, University of British Columbia, 1979.

2 On the early mine at Fort Rupert and the role of the Hudson's Bay Company, see Bowen, *Three Dollar Dreams;* Newsome, *The Coal Coast;* H. Keith Ralston, "Miners and Managers: The Organization of Coal Production on Vancouver Island by the Hudson's Bay Company, 1848-1862," in *The Company on the Coast,* ed. E. Blanche Norcross (Nanaimo: Nanaimo Historical Society, 1983). See also Gallacher, "Men, Money, Machines," and William Burrill, "Class Conflict and Colonialism: The Coal Miners of Vancouver Island during the Hudson's Bay Company Era, 1848-1862," MA thesis, University of Victoria, 1987.

3 See also Bowen, *Three Dollar Dreams,* 291ff. On Dunsmuir's activities in the Comox Valley, see Richard Somerset Mackie, *The Wilderness Profound: Victorian Life on the Gulf of Georgia* (Victoria: Sono Nis Press, 1995).

4 Reksten, *The Dunsmuir Saga,* 61.

5 Ibid., 58-59.

6 The grant did not include land that had been pre-empted prior to 1884.

7 *Nanaimo Free Press,* 29 May 1891, quoted from Allen Seager and Adele Perry, "Mining the Connections: Class, Ethnicity, and Gender in Nanaimo, British Columbia, 1891," *Histoire sociale/Social History* 30 (1997): 63, 58.

8 R.A.J. McDonald, *Making Vancouver, 1863-1913* (Vancouver: UBC Press, 1996), 75.

9 Craigdarroch perhaps best symbolizes his feudal attitude towards his employees. His final rite of passage towards baronial status, a knighthood, was denied because of his alleged "'open and advised' enthusiasm for an American union [which] was 'disloyal to the Queen' and 'a violation of allegiance and oath of office'": Reksten, *The Dunsmuir Saga,* 90.

10 Ibid., 50-53. Quote from 53.

11 Reksten, *The Dunsmuir Saga,* 62-63. Dunsmuir received 226 of 424 ballots cast.

12 McDonald, *Making Vancouver,* 75. For a consenting but less understated view, see also Martin Robin's appropriately entitled chapter "An Early Barbecue," in *The Rush for Spoils: The Company Province, 1871-1933* (Toronto: McClelland and Stewart, 1972); and Margaret A. Ormsby's chapter "The Great Potlatch," in *British Columbia: A History* (Toronto: Macmillan, 1958). See also Jean Barman, *The West Beyond the West: A History of British Columbia* (Toronto: University of Toronto Press, 1991).

13 Quoted in Reksten, *The Dunsmuir Saga,* 164.

14 *Report of the Royal Commission on Industrial Disputes in the Province of British Columbia* (Ottawa: Department of Labour, 1903), Minutes of Evidence, 379. Hereafter cited as *Royal Commission, 1903.*

15 See Seager and Perry, "Mining the Connections," 58.

16 Quoted in Bowen, *Three Dollar Dreams,* 153. Although they never recognized a union, they did hire striking miners back.

17 Price V. Fishback, *Soft Coal, Hard Choices: The Economic Welfare of Bituminous Coal Miners, 1890-1930* (New York: Oxford University Press, 1993); Price Fishback, "The Economics of Company Housing: Theoretical and Historical Perspectives from the Coal Fields," *Journal of Law, Economics, and Organization* 8 (1992): 346-65; Price Fishback, "The Miner's Work Environment: Safety and Company Towns in the Early 1900s," in *The United Mine Workers of America: A Model of Industrial Solidarity,* ed. John H.M. Laslett (University Park, PA: Pennsylvania State University Press, 1996), 201-23; Price Fishback and Dieter Lauszus, "The Quality of Services in Company Towns: Sanitation in Coal Towns During the 1920s," *Journal of Economic History* 49 (1989): 125-44; Price Fishback, "Did Miners 'Owe Their Souls to the Company Store'? Theory and Evidence from the Early 1900s," *Journal of Economic History* 46 (1986): 1011-29.

18 See, for instance, David Alan Corbin, *Life, Work, and Rebellion in the Coal Fields,* 116ff.; Priscilla Long, *Where the Sun Never Shines: A History of America's Bloody Coal Industry* (New York: Paragon House, 1989), 78ff.; and Mildred Allen Beik, *The Miners of Windber: The Struggles of New Immigrants for Unionization, 1890s-1930s* (University Park, PA: Pennsylvania State University Press, 1996).

19 Bowen, *Three Dollar Dreams,* 141-42.

20 See, for instance, Fishback, "The Miner's Work Environment," 214.

21 Bowen, *Three Dollar Dreams,* 142.

22 During the drive to unionize miners at Cumberland in 1890, one of the main issues was the high prices at the company store. See Eugene Forsey, *Trade Unions in Canada, 1812-1902* (Toronto: University of Toronto Press, 1982), 364. On housing and store profits, see Long, *Where the Sun Never Shines,* 79.

23 Reksten, *The Dunsmuir Saga,* 37. See also John Douglas Belshaw, "The Standard of Living of British Miners on Vancouver Island, 1848-1900," *BC Studies* 84 (1989-90): 53f.
24 See, for instance, *Report of the Select Committee on the Wellington Strike* (BC Sessional Papers, 1891), 243, and *Royal Commission, 1903,* 38.
25 See Bowen, *Three Dollar Dreams,* 166ff. and 339ff., and Jeremy Mouat, "The Politics of Coal: A Study of the Wellington Miners' Strike of 1890-91." *BC Studies* 77 (1988): 3-29. Eviction, however, was not always thorough. To cite one example, John Anderson was on strike for nine months during 1890-91 but still lived in his company house in Wellington: *Report of the Select Committee on the Wellington Strike,* 243.
26 *Royal Commission, 1903,* 461, 479, and 498.
27 In 1899 the average wage for a white hewer at Nanaimo was between $3.00 and $4.50 per diem. At Wellington the average was between $3.00 and $3.50 per diem. See Belshaw, "The Standard of Living," 42.
28 Bowen, *Three Dollar Dreams,* 144; see also Seager and Perry, "Mining the Connections," 71.
29 Mouat, "The Politics of Coal," 9. See also *Report of the Select Committee on the Wellington Strike.*
30 An intense rivalry developed between Nanaimo and Wellington over the competition for market share and the different business philosophies of Dunsmuir and the VCMLC's manager, Samuel Robins. The friction between Robert Dunsmuir and Sons and the VCMLC, especially after 1890, when Robins recognized the Miners' and Mine Labourers' Protective Association (MMLPA), was sometimes palpable. For instance, during the 1890 strike, Dunsmuir tried to harm the economy of Nanaimo, which was dominated by the VCMLC, by offering "a free train ride to Victoria after each payday so Wellington dollars would bypass Nanaimo": Bowen, *Three Dollar Dreams,* 338.
31 Seager and Perry, "Mining the Connections," 66, argue incorrectly, I believe, that unlike the VCMLC in Nanaimo, "the rival Dunsmuirs never established a well-developed 'company town' in the 1870s, 1880s, or 1890s." However, company towns were characterized by their impermanence and "perambulating geography," their dreary and chaotic spatial order, and the social, political, and economic control operators had over the workforce.
32 Frances Macnab, *British Columbia for Settlers: Its Mines, Trade and Agriculture* (London: Chapman and Hall, 1898), cited in Robert McIntosh, *Boys in the Pits: Child Labour in Coal Mines* (Montreal and Kingston: McGill-Queen's University Press, 2000), 113.
33 *Report of the Select Committee on the Wellington Strike,* 249.
34 S.P. Planta, "The Coal Fields of Vancouver Island, British Columbia," *Canadian Mining Manual* (1893): 294.
35 Quoted in Belshaw, "The Standard of Living," 58, n. 87.
36 Robins supervised the clearing and preparation of the land, known as "the farm," for settlement. Tile drainage was installed, peat was applied to gravelly areas, and fences, roads, and paths were made, including "lovers' walks": H. Mortimer Lamb, "The Coal Industry of Vancouver Island, British Columbia," *British Columbia Mining Record* 4 (1898): 24-26. Robins employed Chinese labour to clear about 700 acres of land. The leaseholders, employing Chinese, had also cleared between 600 and 700 acres. See *Report of the Royal Commission on Chinese and Japanese Immigration* (Ottawa: S.E. Dawson, 1902), 73.
37 Planta, "The Coal Fields of Vancouver Island," 291.
38 *Report of the Royal Commission on Chinese and Japanese Immigration,* 73.

39 According to Robins: "The company leases to miners with the option of purchase, so they can do what they please. Most of the miners who have arrived at marriageable age are married. A great many own their own homes. Large numbers may be considered permanent residents": *Report of the Royal Commission on Chinese and Japanese Immigration,* 74. The five-acre land scheme also enabled the miner and his family to supplement both their diet and income, as many of the lots were transformed into small farms. See Ben Moffat, "A Community of Working Men: The Residential Environment of Early Nanaimo, BC, 1875-1891," MA thesis, University of British Columbia, 1982.

40 Mouat, "The Politics of Coal," 6-7. See also *Royal Commission, 1903,* Minutes of Evidence, 299, for testimony about the five-acre lots. Miners leased them for a period of twenty-one years with the option of buying them after ten years. Rents were $2.50 per year for the first two years, rising to $2.50 per year per acre for the next three years. For the remaining years, the rent increased to $50 per year for the entire lot. At the end of ten years, the miners could buy the lot for between $125 and $200, depending on its location. See also Belshaw, "The Standard of Living," 52-53. City lots, in contrast, sold for between $300 and $500.

41 Allen Seager, "Miners' Struggles in Western Canada: Class, Community, and the Labour Movement, 1890-1930," in *Class, Community and the Labour Movement: Wales and Canada, 1850-1930,* ed. Deian R. Hopkin and Gregory S. Kealey (Oxford: Oxford University Press, 1989), 168.

42 Gallacher describes in detail the inefficiencies of the VCMLC. Technical and managerial problems meant that the firm could not maintain "a constant rate of production." The Wellington collieries, in contrast, were much more efficient. "Between 1879-83, the Dunsmuir collieries' annual output more than doubled while that of the VCMLC dropped by two-thirds": Daniel T. Gallacher, "Men, Money, Machines: Studies Comparing Colliery Operations and Factors of Production in British Columbia's Coal Industry to 1891," PhD thesis, University of British Columbia, 1979, 184.

43 Bowen, *Three Dollar Dreams,* 224 and 226. Reksten, *The Dunsmuir Saga,* 49.

44 For a copy of the agreement, see *Royal Commission, 1903,* Exhibit 9b, 752 and 300, for Robins's view on the WFM. See also John Belshaw, "The British Collier in British Columbia: Another Archetype Reconsidered," *Labour/Le Travail* 34 (1994): 21, n. 36. At the same time, however, Robins apparently dismissed James Hawthornthwaite, future socialist MLA, from his job at the VCMLC in early 1902 because of his criticism of Liberal MP Ralph Smith's moderate line. See Thomas Robert Loosmore, "The British Columbia Labor Movement and Political Action, 1879-1906," MA thesis, University of British Columbia, 1954, 176. Smith also enjoyed good relations with James Dunsmuir, leading many miners to suggest that he was a "company man."

45 See Robins's testimony in *Royal Commission, 1903,* Minutes of Evidence, 294-304; and Belshaw, "The British Collier in British Columbia," 18-20.

46 *Royal Commission, 1903,* Report, 64, and Minutes of Evidence, 295-96.

47 Rumours circulated in early 1890 that the company was going to roll back the wages of pushers to the previous level. In February a committee went to discuss this matter with the superintendent, Bryden, who denied the truth of the rumour. In May the workers struck. See Mouat, "The Politics of Coal," and *Report of the Select Committee on the Wellington Strike,* 259.

48 Alan Grove and Ross Lambertson, "Pawns of the Powerful: The Politics of Litigation in the Union Colliery Case," *BC Studies* 103 (Fall 1994): 10-11.

49 *Royal Commission, 1903,* Minutes of Evidence, 298. Robins and Nanaimo miners often claimed that the Wellington collieries had "unfair" competitive advantages. Was this tactic used to keep Nanaimo miners "cooperative"?
50 Belshaw, "The British Collier in British Columbia," 29, n. 65.
51 As Seager and Perry, "Mining the Connections," 63, point out, the VCMLC proved the "commercial viability of union labour. In all but two years between 1888 and 1901 the Vancouver Coal Company outproduced the Dunsmuirs (exclusive of the latter's Cumberland mines)."
52 Seager, "Miners' Struggles in Western Canada," 168.
53 BC Archives, MS-0436, A.F. Buckham Personal and Professional Papers, 1858-1968, Vol. 146, File 10. See also Bowen, *Three Dollar Dreams,* 360. Robins apparently agreed to sell the property on the condition that Dunsmuir employ no Chinese labour underground.
54 *Royal Commission, 1903,* Minutes of Evidence, 268.
55 BC Archives, A.F. Buckham Personal and Professional Papers, Vol. 146, File 10. For purchase of land at Oyster Harbour, see *Royal Commission, 1903,* Minutes of Evidence, 242. Two men, a Mr. Kemp and a Mr. Nicholson, had bought two parcels of 160 acres each from the E&N. Dunsmuir subsequently bought them back.
56 See Richard Goodacre, *Dunsmuir's Dream* (Victoria: Porcépic Books, 1991), 18. See also BC Archives, MS-0783, John Wood Coburn, "Early History of the Town of Ladysmith, British Columbia."
57 BC Archives, A.F. Buckham Personal and Professional Papers, Vol. 146, File 10.
58 *Annual Report of the Minister of Mines of British Columbia,* 1897, 626; 1898, 1181. The main tunnel was to be a mile long. At the time of the inspector's report in 1898, the men had driven the tunnel 2,624 feet in the direction of the east level.
59 BC Archives, A.F. Buckham Personal and Professional Papers, Vol. 146, File 10.
60 As John Bryden testified, "If they could have succeeded in getting Departure Bay they would not have gone to Ladysmith": *Royal Commission, 1903,* Testimony of John Bryden, 263.
61 *Annual Report of the Minister of Mines,* 1901, 962; BC Archives, A.F. Buckham Personal and Professional Papers, Vol. 146, File 10; Goodacre, *Dunsmuir's Dream,* 20; Bowen, *Three Dollar Dreams,* 363.
62 Dunsmuir renamed the town Ladysmith following the relief of the South African city of the same name during the Boer War, and is an example of how strong patriotism could be, even in the most remote corner of Empire. This imperial identity is also reflected in the names of the major east-west streets, which were named after Boer War generals.
63 *Nanaimo Free Press,* 16 March 1903. Not surprisingly, the *Ladysmith Leader* accused the *Free Press* of slander.
64 Most obviously, Goodacre's book, *Dunsmuir's Dream.*
65 *Royal Commission, 1903,* Minutes of Evidence, 270.
66 BC Archives, A.F. Buckham Personal and Professional Papers, Vol. 1, File 11.
67 Ibid., File 10.
68 *Royal Commission, 1903,* 47 and 264.
69 Ibid., 50.
70 Ibid., 243. The selling of lots led to the incorporation of Wellington and to a mini construction boom. See Bowen, *Three Dollar Dreams,* 355.
71 *Royal Commission, 1903,* 242 and 380.

72 Ibid., 82. BC Archives, A.F. Buckham Personal and Professional Papers, Vol. 146, File 14, Letter from James Dunsmuir to Jacob Mylhylmakie, 5 October 1900. The spelling of this name differs in the sources.

73 *Royal Commission, 1903,* 86 and 83.

74 Ibid., 46, 83, 356.

75 Ibid., 51.

76 Ibid., 46; Testimony of Thomas Isherwood, 79.

77 On the costs of the move, *Royal Commission, 1903,* 48; on wages, 762. "All the company would do was to run the timber down on its railway from one place to the other": *Victoria Daily Times,* 9 May 1903; see also *Royal Commission, 1903,* 79.

78 Goodacre repeats this, adding that "a similar offer for businesses wishing to establish themselves in the new town" was also made: *Dunsmuir's Dream,* 25.

79 *Royal Commission, 1903,* Minutes of Evidence, 244-45. *Victoria Daily Colonist,* 21 April 1901. Apparently the cost of the lots with the cottages ran as high as $500. See Belshaw, "The Standard of Living," 54.

80 *Royal Commission, 1903,* 48.

81 Ibid., 243.

82 The "men" Dunsmuir was referring to were union leaders, who he felt told the miners what to think, "eat, drink and avoid." Still, it is clear that he believed that the workers were essentially children. See *Royal Commission, 1903,* 381.

83 Ibid., 76-77 and 243. See also Bowen, *Three Dollar Dreams,* 360-61. Bramley sold 50 acres to the colliery with the proviso that a roadway be built, as the property was in the centre of the town site. Dunsmuir had a high fence built around it, however, in effect cutting off the community. This caused much friction, not least because children had to walk one mile around the fence to get to school, which was situated at the other side of the village: *Victoria Daily Times,* 9 May 1903.

84 *Royal Commission, 1903,* 243.

85 Ibid., 380.

86 Ibid., 356.

87 Quoted in Reksten, *The Dunsmuir Saga,* 176. In sharp contrast, King thought that Robins was "a gentleman of more than exceptional quality, I never remember meeting any employer ... who impressed me as being more genuinely humane, more truly considerate and courteous and more honorable in his views of life and dealings with men": quoted in Mouat, "The Politics of Coal," 6-7.

**Chapter 2: A Town of Merry Hearts**

1 On the economic history of British Columbia during this period, see especially Allen Seager, "The Resource Economy, 1871-1921," in *The Pacific Province,* ed. Hugh J.M. Johnston (Vancouver: Douglas and McIntyre, 1996), 205-52; Robert McDonald, "Victoria, Vancouver, and the Economic Development of British Columbia, 1886-1914," in *British Columbia Historical Readings,* ed. W. Peter Ward and Robert A.J. McDonald (Vancouver: Douglas and McIntyre, 1981), 369-95; R.A.J. McDonald, *Making Vancouver, 1863-1913* (Vancouver: UBC Press, 1996); D.G. Paterson, "European Finance Capital and British Columbia: An Essay on the Role of the Regional Entrepreneur," in Ward and McDonald, *British Columbia Historical Readings,* 328-42; David J. Reid, "Company Mergers in the Fraser River Salmon Canning Industry," in Ward and McDonald, *British Columbia Historical Readings,* 306-27; Jeremy Mouat, *Roaring Days: Rossland's Mines and the History of British Columbia* (Vancouver: UBC Press, 1995); James Conley, "Relations

of Production and Collective Action in the Salmon Fishery, 1900-1925," in *Workers, Capital, and the State in British Columbia,* ed. Rennie Warburton and David Coburn (Vancouver: UBC Press, 1988), 86-116.

2 See McDonald, "Victoria, Vancouver, and the Economic Development of British Columbia," and *Making Vancouver;* see also John Lutz, "Losing Steam: The Boiler and Engine Industry as an Index of British Columbia's Deindustrialization, 1880-1915," Canadian Historical Association *Historical Papers* (1988): 182-202.

3 McDonald, "Victoria, Vancouver, and the Economic Development of British Columbia," 381 and 371.

4 Andrew Yarmie, "The State and Employers' Associations in British Columbia: 1900-1932," *Labour/Le Travail* 45 (2000): 61-62.

5 *Annual Report of the Minister of Mines of British Columbia,* 1898, K962 and K971. On the general economic recession, see Margaret A. Ormsby, *British Columbia: A History* (Toronto: Macmillan, 1958), 312-14.

6 McDonald, "Victoria, Vancouver, and the Economic Development of British Columbia," 381; Seager, "The Resource Economy," 217; Alan Wargo, "The Great Coal Strike: The Vancouver Island Coal Miners' Strike, 1912-1914," BA thesis, University of British Columbia, 1962, 4ff. Allen Seager and Adele Perry, "Mining the Connections: Class, Ethnicity, and Gender in Nanaimo, British Columbia, 1891," *Histoire sociale/Social History* 30 (1997): 62, write that in 1891 the industry had not *yet* entered a "crisis of dependent development," and that "opportunity, not crisis," was the "local economic keynote."

7 *Annual Report of the Minister of Mines,* 1898, 1166; 1912, K221.

8 BC Archives, Add. Mss. 780, "Report on the Properties of the Wellington Colliery Co.," 1910, 20-22. This report was prepared for Mackenzie and Mann by accountants in Britain prior to their purchase of Dunsmuir's interests. The report indicates that the company made a profit of $1.023 per ton.

9 Seager, "The Resource Economy," 217.

10 *Annual Report of the Minister of Mines,* 1904, H19; 1911, K170. See also Paul Phillips, "The Underground Economy: The Mining Frontier to 1920," in Warburton and Coburn, *Workers, Capital, and the State in British Columbia,* 49, for claims that the Crow's Nest mines displaced the production of coke from the Comox collieries. However, production of coke at Comox made up a fraction of total production. Between 1895 and 1897, only 19,396 tons were produced. See *Report of the Royal Commission re: Coal in British Columbia* (Victoria: King's Printer, 1914), 9. According to the provincial mineralogist, the two fields "are in no way competitors in the market, their markets being quite separate and ruled by completely different conditions": *Annual Report of the Minister of Mines,* 1912, K243. Indeed, the main competition for the Kootenay producers came from Alberta.

11 "Report on the Properties of the Wellington Colliery Co.," 22.

12 *Annual Report of the Minister of Mines,* 1904, H18.

13 *Report of the Royal Commission on Industrial Disputes in the Province of British Columbia* (Ottawa: Department of Labour, 1903), 62. Hereafter cited as *Royal Commission, 1903.*

14 There was a slight increase in imports of Australian coal from 178,563 tons in 1900 to 276,186 in 1903, but a decline in other sources. In 1900 Washington mines sold 668,642 tons to California. By 1903 this had dropped to 484,645 tons. See *Annual Report of the Minister of Mines,* 1901, 954; 1904, H220. The 1904 report also stressed that the market had been "invaded by local fuel in the form of oil": H18. See also *Report of Royal Commissioner on Coal Mining Disputes on Vancouver Island* (Ottawa: Government Printing Bureau, 1913), 7, and Seager, "The Resource Economy," 217.

15 *Annual Report of the Minister of Mines,* 1912, K221.
16 Ibid., 1906, H19 and H220.
17 Ibid., 1911, K170.
18 There was a 48.8 percent increase of total provincial production between 1908 and 1910. The increase was roughly split between the two producing regions: ibid., 1911, K171.
19 *Victoria Daily Colonist,* 20 January 1901.
20 Ibid., 21 April 1901.
21 Ibid., 20 January 1901 and 21 April 1901.
22 BC Archives, GR-0429, Box 6, File 3, 3816/00.
23 Ibid., Box 12, File 2, 1956/05.
24 *Ladysmith Chronicle,* 14 December 1910. Likewise with the arrival of the Royal Bank.
25 Richard Goodacre, *Dunsmuir's Dream* (Victoria: Porcépic Books, 1991), 29.
26 *Ladysmith Chronicle,* 14 December 1910.
27 For example, Seager, "The Resource Economy," 217.
28 *Ladysmith Chronicle,* 14 December 1910.
29 "A Bit of England in the New World: Duncans, Ladysmith and Nanaimo, Vancouver Island," *The Coast* 13 (May 1907): 319.
30 *Ladysmith Chronicle,* 14 December 1910.
31 See especially Goodacre, *Dunsmuir's Dream,* 25 and 33.
32 Ibid., 31.
33 Isabelle Davis, "Forty-Ninth Parallel City: An Economic History of Ladysmith," BA thesis, University of British Columbia, 1953, 21.
34 *Ladysmith Daily Ledger,* 4 January 1905. See also the "Municipal Notice" distributed for the 1907 civic election. I would like to thank Mr. Ray Knight for providing me with a copy of this document. BC Archives, MS-0783, John Wood Coburn Papers, "Early History of the Town of Ladysmith." In the first council, for example, the mayor was John Coburn, businessman. The aldermen included D. Nicholson, contractor; George Haworth, the manager of the opera house; William Beveridge, hotel keeper; Murdoch Matheson, carpenter; and Henry Blair, miner.
35 Gordon Hak, "The Socialist and Labourist Impulse in Small-Town British Columbia: Port Alberni and Prince George, 1911-33," *Canadian Historical Review* 70 (1989): 542. The role of the middle class in the development of the frontier resource community of Dawson City is elaborated in great depth in Charlene Porsild, *Gamblers and Dreamers: Women, Men, and Community in the Klondike* (Vancouver: UBC Press, 1998).
36 Viola Johnson-Cull, comp., *Chronicle of Ladysmith and District* ([Ladysmith]: Ladysmith New Horizons Historical Society, 1980), 262-63.
37 John Belshaw, "The British Collier in British Columbia: Another Archetype Reconsidered," *Labour/Le Travail* 34 (1994): 22. Belshaw also discusses the concept of "labour aristocracy." See also Yarmie, "The State and Employers' Associations," 65. On this and the importance of mobility, status, identity, and labourism, see Mark Leier, *Red Flags and Red Tape: The Making of a Labour Bureaucracy* (Toronto: University of Toronto Press, 1995), 92-107; and McDonald, *Making Vancouver,* 62-119. Their arguments are conveniently summarized in H.V. Nelles, "Horses of a Shared Colour: Interpreting Class and Identity in Turn-of-the-Century Vancouver," Review Essay, *Labour/Le Travail* 40 (1997): 269-75.
38 Lynne Bowen, "Friendly Societies in Nanaimo: The British Tradition of Self-Help in a Canadian Coal-Mining Community," *BC Studies* 118 (1998): 74-76, emphasizes how early membership in Nanaimo's friendly societies crossed class lines, but says that by

1897 "new membership came almost exclusively from the working class." She suggests that the reason for this "may be that by the late 1890s the middle class found sufficient social activities outside the self-help movement to make lodge membership less attractive."

39 Lynne Bowen, *Three Dollar Dreams* (Lantzville, BC: Oolichan Books, 1987), 336, describes how Nanaimo businessmen supported the miners during the 1890 strike. On attitudes towards monopolists, see Leier, *Red Flags and Red Tape,* 93-94, and McDonald, *Making Vancouver,* 75.

40 See, for instance, Belshaw, "The British Collier"; British miners were also the majority in Pennsylvania. See Priscilla Long, *Where the Sun Never Shines: A History of America's Bloody Coal Industry* (New York: Paragon House, 1989), 55ff., and John H.M. Laslett, "British Immigrant Colliers, and the Origins and Early Development of the UMWA," in *The United Mine Workers of America: A Model of Industrial Solidarity,* ed. John H.M. Laslett (University Park, PA: Pennsylvania State University Press, 1996), 29-50. As Long points out (56), however, the British miners did not necessarily view themselves as a single people: "The Welsh, Scottish, and English felt strong separate national identities." In Streator, Illinois, in the 1870s, each nationality had its society: "The English have the St. George's, the Scotch the St. Andrew's, the Welsh the St. David's."

41 See also Seager and Perry, "Mining the Connections," 68.

42 *Census of Canada,* 1911, 170-71.

43 The 1911 census indicates that there were six African Americans living in Ladysmith, although the number was probably slightly higher. See Johnson-Cull, *Chronicle of Ladysmith,* 308.

44 *Souvenir of the 50th Anniversary of the Corporation of the City of Ladysmith, 1904-54* (Ladysmith: Ladysmith Chamber of Commerce, 1954), 19.

45 *Ladysmith Chronicle,* 14 December 1910.

46 Ibid.

47 Goodacre writes that the First Nations population "played no part in the development of Ladysmith," but then goes on to say that "some worked in the laying of track for the railways, others found limited employment at the wharves sacking coal": *Dunsmuir's Dream,* 43.

48 Cull, *Chronicle of Ladysmith,* 294-96.

49 John Lutz, "After the Fur Trade: The Aboriginal Labouring Class of British Columbia, 1849-1890," *Journal of the Canadian Historical Association* 2 (1992): 70.

50 Bowen, *Three Dollar Dreams,* 74-75.

51 John Douglas Belshaw, "Mining Technique and Social Division on Vancouver Island, 1848-1900," *British Journal of Canadian Studies* 1 (1986): 53-54, and Bowen, *Three Dollar Dreams,* 59-61, 113.

52 Johnson-Cull, *Chronicle of Ladysmith,* 295-97. In 1890 white miners could earn between $3.00 and $4.50 per day, while Chinese mine labourers earned roughly between $1.00 and $1.50 per day. Johnson-Cull also comments upon the sexual division of labour in the First Nations community. Women were responsible for the household economy – making clothing, housekeeping, food preparation and preservation, knitting and weaving – and the instruction of girls. Men, in contrast, hunted, fished, built canoes and longhouses, and instructed the boys.

53 Ethnic distinctions are discussed in Allen Seager, "Miners' Struggles in Western Canada: Class, Community, and the Labour Movement, 1890-1930," in *Class, Community and the Labour Movement: Wales and Canada, 1850-1930,* ed. Deian R. Hopkin and Gregory S. Kealey (Oxford: Oxford University Press, 1989), 169.

54 Johnson-Cull, *Chronicle of Ladysmith,* 309-13. See also Goodacre, *Dunsmuir's Dream,* 41. Likewise, Northfield was known as Belgian Town because of the large number of Belgian miners there. See Bowen, *Three Dollar Dreams,* 284.

55 On the Croatian community, see the important articles by Zelimir B. Juricic, "Croatians Killed in Ladysmith Mine Blast," *BC Historical News* 26 (Winter 1992-93): 20-23; and "Croatians Enlivened Mining Towns," *BC Historical News* 27 (Summer 1994): 22-25.

56 *Report of the Select Committee on the Wellington Strike* (BC Sessional Papers, 1891), 283. See also Jeremy Mouat, "The Politics of Coal: A Study of the Wellington Miners' Strike of 1890-91," *BC Studies* 77 (1988): 9.

57 Juricic, "Croatians Killed in Ladysmith Mine Blast," 22.

58 BC Archives, GR-0429, Box 19, Pinkerton Reports, 29 August 1913.

59 According to *Henderson's Gazetteer and Directory* of 1910, there were three Chinese shoemakers (Sing Tye, Kee Lan, and Wah Lee), one tailor (G.C. Yuen), one general merchant (Long Hop), one grocer (Mar Sam Sing), and one baker (Hop Lee). See also the story of Mar Sam Sing in Johnson-Cull, *Chronicle of Ladysmith,* 306-7.

60 See, for example, Lynne Bowen, *Boss Whistle* (Lantzville, BC: Oolichan Books, 1982), 73-74, and Johnson-Cull, *Chronicle of Ladysmith,* 344.

61 Johnson-Cull, *Chronicle of Ladysmith,* 339.

62 For a recent criticism of this view, see John Douglas Belshaw, "Cradle to Grave: An Examination of Demographic Behaviour on Two British Columbian Frontiers," *Journal of the Canadian Historical Association* 5 (1994): 41-62.

63 Johnson-Cull, *Chronicle of Ladysmith,* 280.

64 In Nanaimo in 1891 only 29 percent of mine workers "were members of households with resident spouses ... In Wellington more than two thirds (68 per cent) of the miners were lodgers, and only 14 per cent were enumerated as the 'heads' of more-or-less 'conventional' households": Seager and Perry, "Mining the Connections," 70.

65 This was apparently in contrast to loggers, who "generally lived in camp, [were] often transient, and did not necessarily have any permanent connection with the local community, coming into town only during days off": Goodacre, *Dunsmuir's Dream,* 28.

66 The following paragraphs have been adapted from my article "'Stout Ladies and Amazons': Women in the British Columbia Coal-Mining Community of Ladysmith, 1912-14," *BC Studies* 114 (1997): 33-57. Many of the single men were older sons or other relatives who might have lived on their own but still considered themselves to be part of an extended family. The number of school-aged children varied. In 1902 there were 199 pupils (116 boys and 83 girls) at Ladysmith and 242 at Extension (130 boys and 112 girls). In 1906 enrolment at Extension had dropped considerably as people moved to Ladysmith. That year, there were only 86 pupils at Extension (42 boys and 44 girls); at Ladysmith, there were 553 pupils (294 boys and 259 girls). See *Public Schools Report,* 1902, A13; 1906, A15.

67 Belshaw, "Cradle to Grave," 61.

68 As Belshaw claims, "married men with large numbers of dependents had more at stake and thus more to fight for": ibid., 60. This is also the conclusion of Sara Diamond, "A Union Man's Wife: The Ladies' Auxiliary Movement in the IWA, the Lake Cowichan Experience," in *Not Just Pin Money,* ed. Barbara K. Latham and Roberta J. Pazdro (Victoria: Camosun College, 1984), 290, who writes that "the existence of a permanent community fostered the development of unionism and auxiliaries in this single industry town. The possibility of a home and of a wage capable of supporting a wife and children provided an incentive to organize."

69 See Robert McIntosh, "The Boys in the Nova Scotian Coal Mines, 1873-1923," *Acadiensis* 16 (1987): 35-50; and "The Family Economy and Boy Labour, Sydney Mines, Nova Scotia, 1871-1901," *Nova Scotia Historical Review* 13 (1993): 87-100.

70 According to one account, "the miner's wife had little prestige in English class society, but the reputation of being a clean housekeeper was a badge of honour in the community, so most of them kept their houses spotless": Bill Johnstone, *Coal Dust in My Blood: The Autobiography of a Coal Miner* (Victoria: Royal British Columbia Museum, 1993), 6.

71 See also Cynthia Gay Bindocci, "A Comparison of the Roles of Women in Anthracite and Bituminous Mining in Pennsylvania, 1900-20," in *Sozialgeschichte des Bergbaus im 19. und 20. Jahrhundert,* ed. K. Tenfelde (Munich: Beck Verlag, 1992), 682-91. On the concept of the household in the mining community, see Donald L. Hardesty, "The Miner's Domestic Household: Perspectives from the American West," in *Sozialgeschichte des Bergbaus im 19. und 20. Jahrhundert,* 180-96.

72 The issue of washhouses was raised during the Great Strike, and their installation was a recommendation of BC Royal Commission on Labour. However, it was not acted upon. See *Report of the Royal Commission on Labour* (Victoria: King's Printer, 1914), M16. Some of the hotels in Ladysmith built their own washhouses for the miners. See Johnson-Cull, *Chronicle of Ladysmith,* 151.

73 Johnstone, *Coal Dust in My Blood,* 7.

74 See David Frank, "The Miners' Financier: Women in the Cape Breton Coal Towns, 1917," *Atlantis* 8 (1983): 137-43.

75 Elizabeth Jameson, *All that Glitters: Class, Conflict, and Community in Cripple Creek* (Urbana: University of Illinois Press, 1998), 120. See also Nancy M. Forestell, "The Miner's Wife: Working-Class Femininity in a Masculine Context, 1920-1950," in *Gendered Pasts: Historical Essays in Femininity and Masculinity in Canada,* ed. Kathryn McPherson, Cecilia Morgan, and Nancy M. Forestell (Toronto: Oxford University Press, 1999), 148.

76 Though it is not well documented, domestic violence, often influenced by alcohol, appears to have been common. See Robert McIntosh, *Boys in the Pits: Child Labour in Coal Mines* (Montreal and Kingston: McGill-Queen's University Press, 2000), 142.

77 Quoted in Long, *Where the Sun Never Shines,* 42.

78 Belshaw, "Cradle to Grave," 56.

79 McIntosh, *Boys in the Pits,* 106ff.; "The Boys in the Nova Scotian Coal Mines," 41; "The Family Economy and Boy Labour," 87f. See also Bettina Bradbury, "Pigs, Cows, and Boarders: Non-Wage Forms of Survival among Montreal Families, 1861-91," *Labour/Le Travail* 14 (1984): 9-48. This was the case not only in the Canadian context. For Germany, for instance, see Karin Hartewig, *Das unberechenbare Jahrzehnt: Bergarbeiter und ihre Familien im Ruhrgebiet, 1914-1924* (Munich: Beck Verlag, 1993), 189-91.

80 Seager and Perry, "Mining the Connections," 75. See also Dorothy Schwieder, *Black Diamonds: Life and Work in Iowa's Coal Mining Communities, 1895-1925* (Ames: Iowa State University Press, 1983), 86ff., for a discussion of Italian American women.

81 Juricic, "Croatians Enlivened Mining Towns," 22, and Johnson-Cull, *Chronicle of Ladysmith,* 196 and 259. Tenfelde also refers to the ubiquitous *"Bergmannskuh,"* or "miner's cow," a common sight in the Ruhr Valley coal-mining communities: *Sozialgeschichte der Bergarbeiterschaft,* 323.

82 Jameson, *All that Glitters,* 117-18, discusses at length the notions of "true womanhood" and "true manhood."

83 On the impact of injury and illness on the household and the lives of women, see Forestell, "The Miner's Wife," 145-47. Thus, when Ethel Croston's father died in Wellington, leaving three children, her mother became a midwife, a job she continued after moving to Ladysmith: Johnson-Cull, *Chronicle of Ladysmith,* 282. The importance of midwives in mining communities is stressed by Bindocci, "A Comparison of the Roles of Women," 685.

84 Long, *Where the Sun Never Shines,* 42.

85 Schwieder, *Black Diamonds,* 99. So important was their contribution to the survivability of the mining family and to minimizing the risk of poverty that one historian of mining families of the Ruhr Valley in Germany has referred to the existence of a "second female economy" operating outside the strict confines of the household: Karin Hartewig, *Das unberechenbare Jahrzehnt: Bergarbeiter und ihre Familien im Ruhrgebiet, 1914-1924* (Munich: Beck Verlag, 1993), 191.

86 Angela John, *By the Sweat of Their Brow: Women Workers in Victorian Coal Mines* (London: Croom Helm, 1980), suggests that the lightening of the workload through technological improvements influenced the decision to exclude women. In other words, women were useful when cheap beasts of burden were required. On Vancouver Island during the colonial period, First Nations women fulfilled this function. See also McIntosh, *Boys in the Pits,* 93.

87 As one report by mining officials in the German city of Dortmund declared, women's work in the mines "led unmistakably to the ruin of workers' families": cited in Klaus Tenfelde, *Sozialgeschichte der Bergarbeiterschaft an der Ruhr im 19. Jahrhundert* (Bonn-Bad Godesberg: Verlag Neue Gesellschaft, 1977), 271.

88 Robert Colls, *The Pitmen of the Northern Coalfield: Work, Culture, and Protest, 1790-1850* (Manchester: Manchester University Press, 1987), 139.

89 McIntosh, *Boys in the Pits,* 94.

90 Friedrich Engels, *The Condition of the Working Class in England* (London: Penguin Books, 1987), 255.

91 John Benson, *British Coalminers in the Nineteenth Century* (London: Gill and Macmillan, 1980), 30-31.

92 Colls, *The Pitmen of the Northern Coalfield,* 137. See also John, *By the Sweat of Their Brow,* 51-52.

93 See Bowen, *Three Dollar Dreams.* In the mines of Carmaux, France, for instance, women did not work underground but were sometimes employed in the "cokerie" and on the surface. In other mines, such as in Le Nord, Pas-de-Calais, or in Belgium, women sometimes did work underground. See Rolande Trempé, *Les mineurs de Carmaux: 1848-1914,* vol. 1 (Paris: Les éditions ouvrières, 1971), 132-33. Very few women worked in the huge coalfields of the Ruhr Valley. According to a survey from 1874, for instance, only eighteen women were recorded as being employed at the mines, and there is no evidence that they worked underground: Tenfelde, *Sozialgeschichte der Bergarbeiterschaft an der Ruhr,* 271.

94 *Statutes of British Columbia,* 1877.

95 See Johnson-Cull, *Chronicle of Ladysmith,* 324.

96 Seager and Perry, "Mining the Connections," 73.

97 For instance, in South Wales. See David Gilbert, *Class, Community, and Collective Action* (Oxford: Oxford University Press, 1992), 237.

98 For example, the town's first telephone operators, Ethel and Bertha Clay, and Mary and Edith Pannel, were single, as were seven of the nine schoolteachers. See Johnson-Cull,

*Chronicle of Ladysmith,* 255 and 280; and *Ladysmith Chronicle,* 25 August 1909. Quote from Leier, *Red Flags and Red Tape,* 141.

99 *Report of the Royal Commission on Chinese and Japanese Immigration* (Ottawa: S.E. Dawson, 1902), 81. See also John Belshaw, "The Standard of Living of British Miners on Vancouver Island, 1848-1900," *BC Studies* 84 (1989-90): 55.

100 Ian McKay, "The Realm of Uncertainty: The Experience of Work in the Cumberland Coal Mines, 1873-1927," *Acadiensis* 16 (1986): 24.

101 This attitude was not restricted to coal miners. As Leier has demonstrated, unions also propagated this image: "The union instilled 'courage, manhood, independence, fraternity; the love for the good and the true; it lives in the hearts and minds of the toilers, and must live; it will not die'": *Red Flags and Red Tape,* 140. See also Jameson, *All that Glitters,* 125f.

102 See, for instance, Forestell, "The Miner's Wife," 147, and Gilbert, *Class, Community, and Collective Action,* 237.

103 Miners' daughters also tended to marry at an early age and have large families. See Belshaw, "Cradle to Grave," 56 and 50, 54, 61. "Of the twenty-six miners' daughters married in the Presbyterian church at nearby Ladysmith from 1904 to 1915, all but four married miners": ibid., 56.

104 Lynne Bowen, "Towards an Expanded Definition of Oral Testimony: The Coal Miner on Nineteenth-Century Vancouver Island," in *Work, Ethnicity, and Oral History,* ed. Dorothy E. Moore and James H. Morrison (Halifax: International Education Centre, 1988), 144.

105 Ann Schofield, "An 'Army of Amazons': The Language of Protest in a Kansas Mining Community, 1921-22," *American Quarterly* 37 (1985): 688. As Long has written: "She, no less than her husband, worked in the all-pervasive atmosphere of the coal company and its superintendent": *Where the Sun Never Shines,* 43.

106 See Bowen, "Friendly Societies in Nanaimo." The development of an active associational life was of great importance in the process of social modernization, helping to break down the authority and values of the traditional corporatist structure of society, promoting mobility, public welfare, and democratic participation. In many respects, they represent an intermediate stage in the development of liberal capitalism and lose significance only when the state assumes their function.

107 Sangster writes with regard to the CCF: "Auxiliaries supplied a secure niche for socialist women uncomfortable in the male-dominated mainstream of the party, but who nonetheless wished to offer support based on traditional female roles. And their dances, picnics, euchres, and dinners did contribute meaningfully, in a financial and social sense, to the growth of the CCF. Yet, women's auxiliaries were also places where women were channelled and forgotten by party leaders unconscious and uncaring of the need to break down the sexual division of labour in the party and offer women more challenging political roles": Joan Sangster, *Dreams of Equality: Women on the Canadian Left, 1920-1950* (Toronto: McClelland and Stewart, 1989), 225. See also Linda Kealey, *Enlisting Women for the Cause: Women, Labour, and the Left in Canada, 1890-1920* (Toronto: University of Toronto Press, 1998), 79ff., and Bindocci, "A Comparison of the Roles of Women."

108 *Ladysmith Chronicle,* 13 October 1909; Johnson-Cull, *Chronicle of Ladysmith,* 278-79.

109 Nanaimo Community Archives, Oral History Project, Interview 84-6-8, Viola Johnson-Cull.

**Chapter 3: Down in the Dark and Gloomy Dungeons**

1 *Annual Report of the Minister of Mines of British Columbia,* 1912, K234.

2 Roy Church, *The History of the British Coal Industry, Vol. 3, 1830-1913: Victorian Preeminence* (Oxford: Clarendon Press, 1986), 201-15 and 311-85, provides an excellent description and analysis of the work process and mining techniques.

3 For a contemporary account of longwall mining on Vancouver Island, see Alexander Sharp, "Some Notes on the Longwall Method of Mining Coal," *Transactions of the Canadian Mining Institute* 16 (1913): 463-501.

4 *Annual Report of the Minister of Mines,* 1897, 624; 1904, H221; 1912, K233. At the No.1 mine at Nanaimo, both methods were also used. The No. 2 slope was mined by longwall and was undercut by a compressed-air mining machine, while the diagonal slope was mined by hand using the pillar and stall method. See W.J. Dick, *Conservation of Coal in Canada* (Toronto: Bryant Press, 1914), 173.

5 See *Annual Report of the Minister of Mines,* 1891, 584-85. Pillar and stall was also used in these mines. Without proper timbering, longwall could be dangerous. Of particular concern was the "first break." The first break occurred "when a newly opened longwall has advanced to a point where the roof of the excavated area has become so large that it can no longer support itself. At first the weaker shales begin to break and crumble around the timbers, with the thick upper layers of sandstone remaining in position until they too finally begin to bend and break. As the break begins, a creaking of timbers is heard, then posts begin to split under the relentless pressure and slabs of shale begin to fall from the roof. This is followed by whip-like cracks and hollow rumblings as the sandstone is torn apart. Usually a first break will last for twenty hours. The crushing and grinding stops only when the excavated area is completely closed, or when the roof has met the floor": Bill Johnstone, *Coal Dust in My Blood: The Autobiography of a Coal Miner* (Victoria: Royal British Columbia Museum, 1993), 117.

6 Priscilla Long, *Where the Sun Never Shines: A History of America's Bloody Coal Industry* (New York: Paragon House, 1989), 41.

7 See especially John Douglas Belshaw, "Mining Technique and Social Division on Vancouver Island, 1848-1900," *British Journal of Canadian Studies* 1 (1986): 45-65. Robert McIntosh, in *Boys in the Pits: Child Labour in Coal Mines* (Montreal and Kingston: McGill-Queen's University Press, 2000), 81, repeats Belshaw's argument that the presence of a large pool of unskilled Asian labour facilitated the introduction of longwall on Vancouver Island.

8 We must heed McKay's warning that "often somewhat tenuous parallels are drawn between the coming of longwall in mining and mass-production techniques in factories": Ian McKay, "The Realm of Uncertainty: The Experience of Work in the Cumberland Coal Mines, 1873-1927," *Acadiensis* 16 (1986): 15.

9 Church, *History of the British Coal Industry,* 276-77.

10 John H.M. Laslett, "British Immigrant Colliers, and the Origins and Early Development of the UMWA," in *The United Mine Workers of America: A Model of Industrial Solidarity,* ed. John H.M. Laslett (University Park, PA: Pennsylvania State University Press, 1996), 36-37.

11 Keith Dix, *Work Relations in the Coal Industry: The Hand-Loading Era, 1880-1930* (Morgantown, WV: Institute for Labor Studies, 1977), 7-8.

12 On the importance of teamwork, see Johnstone, *Coal Dust in My Blood,* 118. Donald Reid, in *The Miners of Decazeville* (Cambridge, MA: Harvard University Press, 1985),

163, writes that even after the introduction of longwall mining in the Aubin Basin of France in the 1930s, "the crew remained a fundamental form of work organization."

13 Church, *History of the British Coal Industry,* 338.

14 The strike lasted from 11 February to 25 February 1903. Admittedly, this was a secondary concern. The main reason for the strike was that Western Fuel eliminated a twenty-five-cent allowance for using safety lamps when it introduced a new and improved safety lamp. The company eventually gave in to worker demands. See *Nanaimo Free Press,* 25 February 1903, and *Report of the Royal Commission on Industrial Disputes in the Province of British Columbia* (Ottawa: Department of Labour, 1903), 343. Hereafter cited as *Royal Commission, 1903.*

15 Belshaw, "Mining Technique," 49.

16 On the use of longwall in the United States, see Long, *Where the Sun Never Shines,* 40-41. Pillar and stall remained the preferred system in Nova Scotia until the 1920s and 1930s. See Robert McIntosh, "The Boys in the Nova Scotian Coal Mines, 1873-1923," *Acadiensis* 16 (1987): 37. An explicit assumption in Belshaw's work is that longwall is a quintessentially modern process, whereas pillar and stall was somehow old-fashioned or anachronistic. Although longwall prevails in underground mining today, pillar and stall is still used. The ill-fated Westray mine in Pictou County, Nova Scotia, to cite just one example, was a pillar and stall mine. The decision to employ this technique was made, as in the nineteenth century, because of the faulty nature of the seam. See *The Westray Story: A Predictable Path to Disaster,* Report of the Westray Mine Public Inquiry, vol. 1 (Halifax: Province of Nova Scotia, 1997), 355ff. See also the comments in *Coal in British Columbia: A Technical Appraisal* (Victoria: The Coal Task Force, 1976), 34. In discussing contemporary conditions, the report states that "as far as western Canada is concerned, the number of people with experience of longwall mining is very limited and, of those who have, none have had successful operating experience under western Canadian conditions."

17 Belshaw argues in "Mining Technique," 48-49, that during the late 1870s, when Bryden was manager at the VCMLC in Nanaimo, he "maintained in his letters to his employers in London that the longwall method was best suited for the geology of the Nanaimo mines," although he "preferred pillar-and-stall." However, Bryden does not maintain that it was best suited. Rather, he states in his diaries and letter book that longwall was being successfully used in only one district of No. 1 north level, where the seam was particularly thin ("not more than 2 feet"). This was the only way that it could be profitably mined. Bryden writes: "I do not however claim that there is any advantage in the getting of this coal by the Longwall System over the Pillar and Stall system, where the coal is of sufficient height to work by the latter method, as this coal is of a very dead nature and is but little affected by the subsidence of the roof, therefore, it does not assist in the getting of the coal to the same extent that longwall work generally does. But in the present case the coal is too thin to work to advantage in any other method": John Bryden, Diary and Letter Book, 1878-80, BC Archives, 14 January 1878, 23 January 1878, 26 September 1878.

18 The case of East Wellington in 1892 is similar to the situation of small mines described by McKay: "Longwall and machine mining preserved the tiny collieries of the Joggins field and allowed them to persist as atypical islands of petit-bourgeois ownership in a sea of monopoly capital": McKay, "The Realm of Uncertainty," 16. However, the East Wellington Colliery Company could not survive on longwall alone, and in 1895 it was purchased by Dunsmuir, who moved the machinery to South Wellington. See Lynne

Bowen, *Three Dollar Dreams* (Lantzville, BC: Oolichan Books, 1987), 281, and S.P. Planta, "The Coal Fields of Vancouver Island, British Columbia," *Canadian Mining Manual* (1893): 295.

19 See *Annual Report of the Minister of Mines,* 1897. In 1892 S.P. Planta wrote in his description of the Island collieries: "The workings, as are all those of the mines of Messrs. Dunsmuir & Sons, are on the pillar and stall method." The new No. 6 pit at Wellington, which reached the seam of coal on 1 May 1889, was "from six to eight feet thick in general, with a few thin places; these can be worked longwall, and will be the only workings of that kind in the colliery, as the plan is that of pillar and stall": Planta, "The Coal Fields of Vancouver Island," 292-94.

20 John Belshaw, "The British Collier in British Columbia: Another Archetype Reconsidered," *Labour/Le Travail* 34 (1994): 26; ibid., "Mining Technique," 57. This is repeated in Delphin A. Muise and Robert G. McIntosh, *Coal Mining in Canada: A Historical and Comparative Overview* (Ottawa: National Museum of Science and Technology, 1996), 46. An incomplete survey of methods used between 1897 and 1902 suggests that pillar and stall predominated. Of the ten mines that operated on average during any given year, Dunsmuir's Wellington No. 5 mine and Union No. 5 used both methods, as did the VCMLC's Protection Island and Northfield mines. The others used pillar and stall.

21 Robert Colls, *The Pitmen of the Northern Coalfield: Work, Culture, and Protest, 1790-1850* (Manchester: Manchester University Press, 1987), 22-23.

22 These are recent rates of extraction. See *The Westray Story,* 359, n. 21. *Coal in British Columbia,* 34, gives a higher extraction rate for longwall. Under "favourable conditions," 85-90 percent of the coal was recovered.

23 Dick, *Conservation of Coal,* 187.

24 Ibid. See Colls, *The Pitmen of the Northern Coalfield,* 20-22. See also Long, *Where the Sun Never Shines,* 40.

25 Dix, *Work Relations,* 72-73. That only the most experienced miners were employed robbing pillars partially explains why so many skilled colliers were killed or injured. See also Long, *Where the Sun Never Shines,* 40.

26 See the comments by the miner John Vedder, *Report of the Select Committee on the Wellington Strike* (BC Sessional Papers, 1891), 319. Especially important was the miner's ability to listen to the roof for signs of collapse.

27 *Nanaimo Free Press,* 27 August 1896.

28 McKay, "The Realm of Uncertainty," 11.

29 *Annual Report of the Minister of Mines,* 1912, K234.

30 H. Mortimer Lamb, "The Coal Industry of Vancouver Island, British Columbia," *British Columbia Mining Record* 4 (1898): 18 and 20.

31 Ibid., 17.

32 Sometimes a miner made use of a device known as a cracket. "This was made of wood and looked like a stool, but built low with a sloping top. He used his cracket when he had to lie on his side to under-cut the seam. By placing it under his arm and shoulder, he could keep his lower shoulder clear of the floor and swing the pick freely": Johnstone, *Coal Dust in My Blood,* 25.

33 In Nova Scotia, the Mining Act of 1873 required the use of iron wedges instead of powder in gassy mines. However, wedging was hated by the miners because it was so time-consuming. See Donald MacLeod, "Colliers, Colliery Safety and Coal Mine Politics: The Nova Scotian Experience, 1873 to 1910," in Canadian Historical Association *Historical Papers* (1983): 233 and 249.

34 The Coal Mines Regulation Act, 1877, *Statutes of British Columbia,* c. 122, 484-85.

35 See also *Report of the Special Commission Appointed to Inquire into the Causes of Explosions in Coal Mines* (BC Sessional Papers, 1903), H290. In the aftermath of the 1887 and 1888 explosions, shotlighters were employed at Wellington; see *Annual Report of the Minister of Mines,* 1890, 385, and 1891, 584. The VCMLC also began to use electric batteries to fire shots: Bowen, *Three Dollar Dreams,* 275. Apparently this practice was not adopted in the Dunsmuir mines. In 1909 this was one of the recommendations made by the inquest investigating the Extension explosion. See *Annual Report of the Minister of Mines,* 1910, K219.

36 *Report of the Royal Commission re: Coal in British Columbia* (Victoria: King's Printer, 1914), 12. I am not sure what Belshaw means when he writes, "Unskilled or 'oncost' workers provided the muscle necessary to deliver the coal to the surface of the mine, or they cleaned the pit roads, worked the ventilation system, or tended to pitponies, but they contributed little else": "The British Collier," 32. Clearly their contribution was to make the mine function.

37 Andrew Bryden, manager of Extension, estimated that it took "four or five years to become a skilled labourer": *Report of the Royal Commission on Chinese and Japanese Immigration* (Ottawa: S.E. Dawson, 1902), 80.

38 On boys in the mines, see McIntosh, *Boys in the Pits,* and his articles "The Boys in the Nova Scotian Mines: 1873-1923," *Acadiensis* 16 (1987): 35-50, and "The Family Economy and Boy Labour, Sydney Mines, Nova Scotia, 1871-1901," *Nova Scotia Historical Review* 13 (1993): 87-100. See also John Belshaw, "The Standard of Living of British Miners on Vancouver Island, 1848-1900," *BC Studies* 84 (1989-90): 45 and 55-56.

39 See McIntosh, *Boys in the Pits,* 44, table 3.1. In 1880, 600 boys constituted 18 percent of the mine labour force in Nova Scotia. The peak year for juvenile employment seems to have been 1908, when 999 boys were recorded, although by then they made up only 7.8 percent of the total workforce.

40 McIntosh, *Boys in the Pits,* 48.

41 Underaged boys did work in the mines despite the existing regulations. For instance, Edward Wilmer, a Belgian miner at Wellington, testified in 1891 that he worked with his boy, who was "under age." His boy received "company wages," or $2.50 per diem: *Report of the Select Committee on the Wellington Strike,* 320. D.T. Gallacher, "Men, Money, Machines: Studies Comparing Colliery Operations and Factors of Production in British Columbia's Coal Industry to 1891," PhD thesis, University of British Columbia, 1979, 223, reports that "between 1850 and 1890, eight of the eleven large strikes were related to wages." Belshaw identifies the lack of opportunities for boys and the concomitant inability of miners to control access to the trade as a cause of industrial unrest. See "The British Collier," 34, and "The Standard of Living," 64.

42 *Report of the Royal Commission on Chinese and Japanese Immigration,* 78.

43 Ibid., 73-74, 81-82; *Royal Commission, 1903,* Testimony of Aaron Barnes, 357. Robins continued to employ the Chinese as labourers above ground in order to remain competitive.

44 *Report of the Royal Commission re: Coal in British Columbia,* 8.

45 Nut coal made up between 51 and 53 percent; pea coal, 13 percent. See Dick, *Conservation of Coal,* 188.

46 David Frank, "Contested Terrain: Workers' Control in the Cape Breton Coal Mines in the 1920s," in *On the Job: Confronting the Labour Process in Canada,* ed. Craig Heron and Robert Storey (Montreal and Kingston: McGill-Queen's University Press, 1986), 104.

47 McKay, "The Realm of Uncertainty," 35.

48 Royden Harrison, ed., *Independent Collier: The Coal Miner as Archetypical Proletarian Reconsidered* (London: Harvester Press, 1978), 5.
49 McKay, "The Realm of Uncertainty," 35.
50 As McIntosh concludes, "no technical innovation dislodged the skilled collier – and the associated division of labour – from his central place in the production of coal until the twentieth century": *Boys in the Pits,* 78.
51 See, for instance, John Norris, "The Vancouver Island Coal Miners, 1912-1914: A Study of an Organizational Strike," *BC Studies* 45 (1980): 56-72, and Allen Seager, "Socialists and Workers: The Western Canadian Coal Miners, 1900-1921," *Labour/Le Travail* 16 (1985): 42.
52 "Lady McKay," as the machine was called by its engineer, Andrew Hunter, began operating in Nanaimo only in 1853, however. See Bowen, *Three Dollar Dreams,* 63-64.
53 *Annual Report of the Minister of Mines,* 1897, 622; 1891, 581. Belshaw, "Mining Technique," 50, also suggests that electrification of the haulage system made mules and ponies redundant. However, mules continued to be used extensively underground at Extension during this time. For a particularly vivid account, see Zelimir B. Juricic, "Mules Miners' Beasts of Burden," *Victoria Times-Colonist,* 19 July 1992.
54 Belshaw, "Mining Technique," 50.
55 Planta, "The Coal Fields of Vancouver Island," 296-97 and 293. See also Lamb, "The Coal Industry of Vancouver Island."
56 See *Annual Report of the Minister of Mines,* 1891, 586-87. See also Planta, "The Coal Fields of Vancouver Island," 296; McIntosh, *Boys in the Pits,* 81-82.
57 *Annual Report of the Minister of Mines,* 1891, 587.
58 Dix, *Work Relations,* 19.
59 Ibid., 20-21.
60 Ibid., 20; *Annual Report of the Minister of Mines,* 1916, K312.
61 *Report of the Royal Commission re: Coal in British Columbia,* 30.
62 Frank, "Contested Terrain," 105.
63 Belshaw, "Mining Technique"; McIntosh, *Boys in the Pits,* 85; Gallacher, "Men, Money, Machines," 224.
64 Quoted from the *Canadian Mining Review,* October 1900, 214. Cited in McIntosh, *Boys in the Pits,* 220, n. 155. Alan J. Wargo, in "The Great Coal Strike: The Vancouver Island Coal Miners' Strike, 1912-14," BA thesis, University of British Columbia, 1962, 8, writes that because of the pinches and rolls, "in many cases it was found impossible to use machinery for cutting."
65 See, for instance, McIntosh, *Boys in the Pits,* 219.
66 Dix, *Work Relations,* 25-27. *Report of the Royal Commission re: Coal in British Columbia,* 31, points out that only when the thinner seams that were mined by longwall were equipped with conveyor systems did production increase. Klaus Tenfelde, in his comprehensive study of mining in Germany's Ruhr Valley, also emphasizes that technical innovation and economic expansion did not greatly change the process of winning coal. See Klaus Tenfelde, *Sozialgeschichte der Bergarbeiterschaft an der Ruhr im 19. Jahrhundert* (Bonn-Bad Godesberg: Verlag Neue Gesellschaft, 1977), 219.
67 Church, *History of the British Coal Industry,* 351, suggests that miners' apprehension about the introduction of mine machines could "be overcome by adjusting payment."
68 See Gallacher, "Men, Money, Machines," 223ff. Layoffs were common during times of slack consumption and recession, although operators often tried to maintain production levels, in the belief that stockpiles could be sold when the economy recovered, in order to avoid layoffs.

69 In May 1867 European miners, fearing for their jobs, struck in response to a proposed wage decrease for Chinese workers. The strike lasted two months and was a victory for the strikers, who won a wage increase. The Chinese were restricted to unskilled labour, above and below ground, and were now hired by European miners as helpers and runners. During the strike, the Victoria newspaper printed the much-quoted statement: "The colliers threaten with violence the first Chinaman who forgets his Celestial origin so far as to descend to the 'bottomless pit' of a coal mine": *Victoria Colonist,* 8 May 1867.

70 Gallacher, "Men, Money, Machines," 211.

71 Patricia Roy, *A White Man's Province: British Columbia Politicians and Chinese and Japanese Immigrants, 1858-1914* (Vancouver: UBC Press, 1989), 40.

72 *Report of the Royal Commission on Chinese and Japanese Immigration,* Testimony of Samuel Robins, 118. Roy, *A White Man's Province,* 60-61, points out that the Victoria-based Anti-Chinese Union merged with the Knights of Labor and that the Knights' membership included "substantial businessmen and property owners as well as workingmen."

73 *Report of the Royal Commission on Chinese and Japanese Immigration,* 119.

74 See the important essay by Alan Grove and Ross Lambertson, "Pawns of the Powerful: The Politics of Litigation in the Union Colliery Case," *BC Studies* 103 (1994): 10, and Roy, *A White Man's Province,* 78; Seager and Perry, "Mining the Connections," 60. Within a year, the percentage of Chinese workers in the mines had dropped from 46.85 in 1887 to 12.81 in 1888.

75 Quoted in Roy, *A White Man's Province,* 138.

76 *Report of the Royal Commission on Chinese and Japanese Immigration,* 76-77. Little also disagreed with the manager at Extension, Andrew Bryden, who preferred Europeans to the Chinese, and was satisfied with the new arrangements. Bryden did not believe that the mine suffered financially. "We would lose $50 a day in employing whites instead of Chinese," he thought. "I don't know whether this would be a serious matter or not": ibid., 80.

77 BC Archives, MS-0436, Vol. 1, File 25, James Dunsmuir to E.P. Bremner, 9 October 1900.

78 *Report of the Royal Commission on Chinese and Japanese Immigration,* 80.

79 Ibid., 77.

80 *Annual Report of the Minister of Mines,* 1902, 1212, 1208.

81 Belshaw, "Mining Technique," 57-58.

82 Belshaw, "The British Collier," 35. Indeed, the opposite may be true. The removal of the Chinese from below ground, with the exception once again of Union, may be seen as a statement of respect for the skilled collier.

83 See, for instance, McIntosh, *Boys in the Pits,* 50, and Belshaw, "The British Collier," 24.

84 *Report of the Special Commission Appointed to Inquire into the Causes of Explosions in Coal Mines,* H293.

85 *Annual Report of the Minister of Mines,* 1910, K209.

86 For instance, there had been no increase in wages at the Wellington collieries since 1874. See Gallacher, "Men, Money, Machines," 227. Belshaw's table in "The Standard of Living," 41, suggests that the base rate remained unchanged between 1875 and 1889.

87 See *Report of the Royal Commission on Chinese and Japanese Immigration,* 81. See also Belshaw's arguments in "The Standard of Living."

88 *Report of the Royal Commission on Chinese and Japanese Immigration,* 111; Roy, *A White Man's Province,* 42. Dunsmuir also maintained that white people refused to do the

same kind of work given to the Chinese: *Report of the Royal Commission on Chinese and Japanese Immigration,* 130.

89 Belshaw, in "The British Collier," 32, writes that "the subsequent introduction of less-skilled workers contributed to the bisection of the underground workforce: one group perceived itself as having skills approximating a 'craft,' the other being regarded as a kind of apprentice class at best." The implication is that this process began only with the introduction of the Chinese. However, this division between skilled and less-skilled workers is as old as coal mining itself. Young boys and men did serve an informal apprenticeship or period of training before becoming hewers.

90 Andrew Markus, *Fear and Hatred: Purifying Australia and California, 1850-1901* (Sydney: Hale and Iremonger, 1979), 255. Quoted in Roy, *A White Man's Province,* ix. The arguments and language used against the Chinese are very reminiscent of those put forth by Friedrich Engels in his discussion of Irish immigrants in England. While he opposes the use of Irish as competition for British labour, he characterizes the Irish as somehow inherently unclean, unhealthy, savage, uncivilized, immoral, and stupid: Engels, *The Condition of the Working Class in England* (London: Penguin Books, 1987), 111-26.

91 *Royal Commission on Chinese and Japanese Immigration,* 1902, 120-21.

92 Ibid., 155-57; *Royal Commission, 1903,* 509.

93 McIntosh, *Boys in the Pits,* 58. Mark Leier also points out that the Chinese were brought into the province in order "to drive down wages and to break trade unions": Leier, *Red Flags and Red Tape: The Making of a Labour Bureaucracy* (Toronto: University of Toronto Press, 1995), 125.

94 Gallacher, "Men, Money, Machines," 225.

95 *Royal Commission, 1903,* 509.

96 Or, indeed, than ideological consistency, as was pointed out by a Conservative MLA, who, to the discomfort of the two Socialist MLAs, declared in the legislature that "the true Socialist, if loyal to his principles, regarded the Chinaman and the Jap as his equals and brother:" cited in Roy, *A White Man's Province,* 254-55.

97 Gillian Creese, "Class, Ethnicity, and Conflict: The Case of Chinese and Japanese Immigrants, 1880-1923," in *Workers, Capital, and the State in British Columbia,* ed. Rennie Warburton and David Coburn (Vancouver: UBC Press, 1988), 73.

98 Rennie Warburton, "The Workingmen's Protective Association, Victoria, BC, 1878: Racism, Intersectionality and Status Politics," *Labour/Le Travail* 43 (1999): 115.

**Chapter 4: Death's Cold Hand**

1 Keserich was one of the many Croatians killed in the disaster. He is mistakenly called "Alex" in the press and official reports. The names were often misspelled. See Zelimir B. Juricic, "Croatians Killed in Ladysmith Mine Blast," *BC Historical News* 26 (Winter 1992-93): 22. Juricic describes William Keserich's body as being the only one to suffer serious burns and disfigurement, when it was apparently that of James. See *Annual Report of the Minister of Mines of British Columbia,* 1910, K209.

2 Wargo's death struck the community hard. He was a much-respected man whose prowess as a hunter made him well known throughout the district. The newspaper wrote: "As a result of an accident some years ago he lost the sight of one eye. He therefore had a special gun made with a crooked stock, so that while he held the gun to his right shoulder he sighted with his left eye. Of course it was a handicap, but Jack still held his own at the traps. He had just received a new gun, and only started work on the morning of the accident after a week's hunting trip": *Nanaimo Free Press,* 7 October 1909.

3 Why Keserich was outside his stall was never determined. A likely reason is that he left after lighting his fuse. If Keserich did fire a blown shot, another plausible scenario, not suggested by the investigators, presents itself. It is possible that the blown shot produced methane that was then ignited by Keserich's lamp. The subsequent explosion would have brought down the first cave-in and expended itself westward (i.e., towards 2½ west level and 3 level). The first cave-in would also have produced methane in the eastern district of 2½ level, which was then ignited at the foot of rooms 20, 21, and 22 (including Wargo's Slant and Peterson's Crosscut, which together formed the dead-end). The second explosion killed the men here and produced the second cave-in. All information comes from *Annual Report of the Minister of Mines,* 1910.

4 Ibid., K216.

5 *Ladysmith Chronicle,* 9 October 1909.

6 D.T. Gallacher, "Men, Money, Machines: Studies Comparing Colliery Operations and Factors of Production in British Columbia's Coal Industry to 1891," PhD thesis, University of British Columbia, 1979, 216 and 222.

7 *Annual Report of the Minister of Mines,* 1911, K178. These rates are based on the total number of men employed in the mining industry. Rates for underground workers only would be higher still. The Kootenay mines were also extremely dangerous, but are not subject to analysis here. In 1910, for instance, the total number of accidents (including fatalities) for the Kootenays was 27 per thousand men employed; for the Coast District, it was 22.59: *Annual Report of the Minister of Mines,* 1911, K232. The death rate varied from year to year. Due to a number of catastrophes in 1901, the death rate was 25.66 per thousand men employed; in 1902, 34.65 per thousand men employed. Averages like this had been common in Britain during the first half of the nineteenth century. Between 1811 and 1815, according to one recent estimate, the death rate in British collieries was close to 10 per thousand. There was considerable improvement over the years. In the pits of northeast England between 1839 and 1845, the death rate was 6.1 per thousand men. See Robert Colls, *The Pitmen of the Northern Coalfield: Work, Culture, and Protest, 1790-1850* (Manchester: Manchester University Press, 1987), 24.

8 Quite likely this is not an accurate assessment of injury rates. These figures indicate only injuries that were reported and recorded; it is possible that the true number was much higher. Accident rates in Britain appear to have been lower than in French and German mines. For instance, in the Ruhr coalfields in 1905, the nonfatal injury rate was 168 per thousand miners (among foreign workers it was 253 per thousand), and French statistics from 1885-87 reveal that 174.8 per thousand men employed were injured per year. Admittedly, these samples are not representative. See statistics, compiled from *Annual Report of the Minister of Mines,* various years. See Stephen Hickey, *Workers in Imperial Germany: The Miners of the Ruhr* (Oxford: Clarendon Press, 1985), 121; Donald Reid, "The Role of Mine Safety in the Development of Working-Class Consciousness and Organization: The Case of the Aubin Coal Basin, 1867-1914," *French Historical Studies* 13 (1981): 98; John Benson, "Mining Safety and Miners' Compensation in Great Britain, 1850-1900," in *Sozialgeschichte des Bergbaus im 19. und 20. Jahrhundert,* ed. Klaus Tenfelde (Munich: Beck Verlag, 1992), 1028 and 1035.

9 The worst catastrophe occurred on 3 May 1887, when two horrific explosions in the Vancouver Coal Mining and Land Company's No. 1 mine on Nanaimo's waterfront decimated the afternoon shift, resulting in the death of 148 men. Nine months later, another explosion, this time at Robert Dunsmuir's Wellington No. 5 pit, took the lives of 77 men. While the death toll was reduced in the 1890s, the new century was ushered

in by two devastating blasts. In 1901 an explosion at Dunsmuir's Cumberland operations killed 64 men. The following year, 125 men lost their lives in an explosion at the Coal Creek colliery at Fernie in the Kootenays.

10 For a recent discussion of methane, see *The Westray Story: A Predictable Path to Disaster,* Report of the Westray Mine Public Inquiry, vol. 1 (Halifax: Province of Nova Scotia, 1997), ch. 8. Quote from 206. Accordingly: "At a concentration of less than 5 per cent, methane will burn only at the points of contact with a surface that is maintained at a sufficiently high temperature. The flame will neither leave that surface nor propagate through the mixture. At a concentration of between 5 and about 15 per cent methane ... flame will propagate spontaneously ... At concentrations above 15 per cent, there will be insufficient oxygen to maintain combustion unless an artificial supply of oxygen has been added to the atmosphere": 278. At 15 percent the flame will be completely extinguished.

11 Ibid., 316.

12 See *Annual Report of the Minister of Mines,* 1890, 385.

13 Lynne Bowen, *Three Dollar Dreams* (Lantzville, BC: Oolichan Books, 1987), 272 and 275-76. Coal dust was also responsible for propagating the explosion in the Courrières mines in 1906. Robert G. Neville, "The Courrières Colliery Disaster, 1906," *Journal of Contemporary History* 13 (1978): 47f., discusses the "coal dust theory," as does Donald MacLeod, "Colliers, Colliery Safety and Coal Mine Politics: The Nova Scotian Experience, 1873 to 1910," in Canadian Historical Association *Historical Papers* (1983): 248. Quote from the *Nanaimo Free Press,* 6 October 1909. This was despite the evidence to the contrary. See especially *Report of the Special Commission Appointed to Inquire into the Causes of Explosions in Coal Mines* (BC Sessional Papers, 1903), H288f. Because the Extension pits were universally damp, coal dust quickly turned into mud. The reason why the explosion was localized was probably because there was little dust to fuel it.

14 A further consequence of the buoyancy of the gas is "methane layering," whereby methane moves upwards and accumulates at high concentrations at the roof. See *The Westray Story,* 283ff.

15 *Annual Report of the Minister of Mines,* 1910, K191.

16 Ibid., 1902, 1203 and 1208; 1903, H272.

17 Priscilla Long, *Where the Sun Never Shines: A History of America's Bloody Coal Industry* (New York: Paragon House, 1989), 32.

18 Benson, "Mining Safety and Miners' Compensation in Britain," 1027.

19 This happened to Joseph Demarks, a miner working in Wellington's No. 5 shaft, to cite just one example. Demarks, unaware that the blast had knocked down the timbers, had returned to his stall in the dust and darkness when about eight tons of coal fell on him. The following year, Ham Fong, a miner at Union, was also killed in his stall. According to the official report, "he was working with his partner when a piece of cap rock fell on his head and knocked his brains out": *Annual Report of the Minister of Mines,* 1901, 973; 1902, H281.

20 Ibid., 1902, 1216.

21 Ibid., 1219.

22 Ibid., 1220; 1901, 975. There were many such stories. Lynne Bowen eloquently relates the death of Sam Lowe in 1883, who apparently mistook the shaft door for a cabin door at East Wellington and fell to his death: Bowen, *Three Dollar Dreams,* 237-38.

23 *Annual Report of the Minister of Mines,* 1902, 1220; 1904, H238; 1905, G292.

24 For a detailed analysis of the geology and mineralogy of Vancouver Island coal, see Gallacher, "Men, Money, Machines," 10-18.

25 H. Mortimer Lamb, "The Coal Industry of Vancouver Island, British Columbia," *British Columbia Mining Record* 4 (1898): 1-2. For a more scientific evaluation of the geology, see Charles H. Clapp, "The Geology of the Nanaimo Coal District," *Transactions of the Canadian Mining Institute* 15 (1912): 334-53.

26 Clapp, "The Geology of the Nanaimo Coal District," 343. See also Robert Strachan, "Coal Mining in British Columbia," *Transactions of the Canadian Institute of Mining and Metallurgy* 26 (1922): 74-75.

27 While irregular seams plagued all the Island operations, at the Union colliery the problem of rolls and pinches was complicated by another particular geological feature. Overlying the main seam at Cumberland was a layer of fireclay that posed additional hazards to the miners, because it was "full of slips which cause the roof to fall in the rooms without cracking and thus, without warning to the miners." This factor increased the accident rate considerably at Cumberland. According to one estimate, three-fifths of the accidents there in 1910 were the result of cave-ins caused by clay slippage. In 1913 the mine inspector reported that 80 percent of the accidents in the No. 2 Union mine were the result of the "friable fireclay" that constituted the overlaying strata. See W.J. Dick, *Conservation of Coal in Canada* (Toronto: Bryant Press, 1914), 189, and *Annual Report of the Minister of Mines,* 1913, K272.

28 *Report of the Royal Commission on Industrial Disputes in the Province of British Columbia* (Ottawa: Department of Labour, 1903), Testimony of Moses Woodburn, 271. Hereafter cited as *Royal Commission, 1903.* According to Woodburn, the seam was "all jumbled up here. Today may be favourable and tomorrow unfavourable. The coal seams are more uniform in the old country."

29 See, for instance, *Report of the Select Committee on the Wellington Strike* (BC Sessional Papers, 1891), 295. John Belshaw, in "The Standard of Living of British Miners on Vancouver Island, 1848-1900," *BC Studies* 84 (1989-90): 40, writes: "It was general practice at Nanaimo to switch from piece-work to daily rates when the mine encountered a shrinking seam."

30 *Report of the Royal Commission on Chinese and Japanese Immigration* (Ottawa: S.E. Dawson, 1902), 80.

31 *Annual Report of the Minister of Mines,* 1891, 590; 1901, 955; 1904, H220.

32 Price V. Fishback, *Soft Coal, Hard Choices: The Economic Welfare of Bituminous Coal Miners, 1890-1930* (New York: Oxford University Press, 1993), 107.

33 This was the decision of the Indiana Supreme Court, the report of which was carried in the *Nanaimo Free Press,* 27 August 1896. As the Indiana Supreme Court explained: "It may be safe to state, however, as a general proposition, that under the usual and ordinary contracts of employment between a master and a servant, whether the engagement be for services in a factory, on a railroad, or in a coal mine, the master undertakes to use reasonable care to see that his machinery is in good condition and repair, and that the place where the servant is to work is free from dangers other than those which are naturally incident upon the work to be performed."

34 *Annual Report of the Minister of Mines,* 1902, 1203; Keith Dix, *Work Relations in the Coal Industry: The Hand-Loading Era, 1880-1930* (Morgantown, WV: Institute for Labor Studies, 1977), 73. For an in-depth analysis of safety and the law, see Eric Tucker, *Administering Danger in the Workplace: The Law and Politics of Occupational Health and Safety Regulation in Ontario, 1850-1914* (Toronto: University of Toronto Press, 1990).

35 The term is from Fishback, *Soft Coal, Hard Choices,* 107.

36 See David Frank, "Contested Terrain: Workers' Control in the Cape Breton Coal Mines in the 1920s," in *On the Job: Confronting the Labour Process in Canada,* ed. Craig Heron and Robert Storey (Montreal and Kingston: McGill-Queen's University Press, 1986).
37 Fishback, *Soft Coal, Hard Choices,* 62.
38 See also Reid, "The Role of Mine Safety," 117, and Hickey, *Workers in Imperial Germany,* 120 and 140ff.
39 *Nanaimo Free Press,* 7 October 1909.
40 Ironically, the system of production bonuses introduced at the Westray mine had the same effect. It rewarded increased production, but encouraged "rushing, taking short-cuts, not taking appropriate breaks." "'[The] bonus system killed the safety'": *The Westray Story,* 186.
41 See MacLeod, "Colliers, Colliery Safety and Coal Mine Politics," 250, and *Report of the Special Commission Appointed to Inquire into the Causes of Explosions in Coal Mines,* H291-92.
42 The safety lamp in use at Nanaimo "was not as good as that of the naked lamp, giving, indeed, owing to the smoke from the oil coating on the glass, a sort of darkness made visible towards the end of a shift. The men consequently worked under increased difficulties": *Nanaimo Free Press,* 5 February 1903.
43 Thus, J. Albertine, a miner at Extension, burned his face and hands in February 1908 when he ignited some gas with his open-flame lamp, although he had been provided with a safety lamp and specifically told to use it. A few months later, Fred Good, a collier at Union, suffered similar injuries when he too refused to use a safety lamp: *Annual Report of the Minister of Mines,* 1909, J236-37.
44 *Report of the Royal Commission re: Coal in British Columbia* (Victoria: King's Printer, 1914), 11. In contrast, miners in the Crow's Nest Pass were paid between 55 and 62 cents per ton. See Belshaw, "The Standard of Living." Immigrant colliers expected higher wages, but this was not always realized. John Belshaw, in "The British Collier in British Columbia: Another Archetype Reconsidered," *Labour/Le Travail* 34 (1994): 35, writes: "What virtually all of the miners did share was the experience of financial hardship and high wage expectations associated with making the move to Vancouver Island."
45 *United Mine Workers Journal,* 20 June 1912.
46 *Report of the Royal Commission re: Coal in British Columbia,* 9.
47 *Report of the Select Committee on the Wellington Strike,* 269.
48 *BC Federationist,* 6 January 1912.
49 Belshaw, "The Standard of Living," 46. In the United States, "in average years while most manufacturing concerns worked 270 to 300 days a year, the typical coal mine was open only 200 to 220 days a year due to overcapacity, seasonal fluctuations, and problems obtaining railroad cars": Fishback, *Soft Coal, Hard Choices,* 79ff.
50 *Report of the Royal Commission re: Coal in British Columbia,* 24. See also Belshaw, "The Standard of Living," 46.
51 This, anyway, was the conclusion of one American report studying employment levels between 1919 and 1939: "We find that a decline of employment is almost invariably accompanied by an increase in the rate of fatal accidents and vice versa," quoted in Dix, *Work Relations,* 102. Absenteeism was also common during fine weather and after payday. See Robert McIntosh, *Boys in the Pits: Child Labour in Coal Mines* (Montreal and Kingston: McGill-Queen's University Press, 2000), 141-42.
52 *Annual Report of the Minister of Mines,* 1911, K177.

53 At Nanaimo it was also customary to switch from piece-rates to daily rates when seams became poor. See Belshaw, "The Standard of Living," 40.
54 *Royal Commission, 1903,* Testimony of Aaron Barnes, 353. See also ibid., Testimony of Samuel Mottishaw, 58-59. For Jones, see *Report of the Select Committee on the Wellington Strike,* 276.
55 See *Royal Commission, 1903,* 353-54. On the question of wages, cavilling, and deficient places, see also Belshaw, "The Standard of Living," 42ff.
56 Gallacher, "Men, Money, Machines," 217, argues that during the 1870s and 1880s, the Nanaimo collieries were more dangerous than the Dunsmuir mines.
57 *Report of the Select Committee on the Wellington Strike,* 271.
58 Ibid., 287.
59 Ibid., 267.
60 *Annual Report of the Minister of Mines,* 1879, 254.
61 Ibid., 252.
62 On the battles waged by the inspectors against the Dunsmuirs, see Bowen, *Three Dollar Dreams,* 305-12.
63 *Report of the Special Commission Appointed to Inquire into the Causes of Explosions in Coal Mines,* H293 and H295. The commission was composed of John Bryden and Tully Boyce.
64 This was recognized by the commission: "It is true some objected to it [examination of blasting by the shotlighter], but we think their objections were based on a not unnatural desire to be unmolested in their work." As Wellington miner John Anderson had declared back in 1891, he was fed up with men "coming into your place and telling you this and that, and directing you how to fix the place." This sentiment was common and was also captured by Robertson in his report on the Extension explosion. It was "*practically impossible* to enforce the [Mining] Act as regards shot-lighting where fuse is employed," he noted, because "where safety-lamps are demanded, it is demoralising to the men to see a shotlighter light a fuse *with an open light.*" See *Report of the Special Commission Appointed to Inquire into the Causes of Explosions in Coal Mines,* H290; *Report of the Select Committee on the Wellington Strike,* 262; *Annual Report of the Minister of Mines,* 1910, K219.
65 The Coal Mines Regulation Act, 1877, *Statutes of British Columbia.*
66 For the following, see John Thomas Keelor, "The Price of Lives and Limbs Lost at Work: The Development of No-Fault Workers' Compensation Legislation in British Columbia, 1910 to 1916," MA thesis, University of Victoria, 1996.
67 Belshaw, "The British Collier," 29.
68 Quoted in Bowen, *Three Dollar Dreams,* 176.
69 See, in particular, Alan Grove and Ross Lambertson, "Pawns of the Powerful: The Politics of Litigation in the Union Colliery Case," *BC Studies* 103 (Fall 1994): 3-31.
70 Ibid., 19.
71 *Report of the Select Committee on the Wellington Strike,* 245.
72 *Western Clarion,* 30 August 1913.
73 See, for instance, Gallacher, "Men, Money, Machines," 221, although he qualifies this by stating that "it took years, however, for this lesson to sink in." Eric Tucker, in *Administering Danger in the Workplace,* 216, writes: "One of the ideological pillars of occupational health and safety regulation, past and present, has been the assumption that labour and capital have a common interest in this area and that 'all stand to gain through collaboration.'"
74 *Nanaimo Free Press,* 18 April 1903.

75 *Annual Report of the Minister of Mines,* 1910, K192.

76 Fishback, *Soft Coal, Hard Choices,* 112.

77 *Royal Commission, 1903,* Testimony of Samuel Robins, 297. See also Lamb, "The Coal Industry of Vancouver Island," 7.

78 See, for instance, Frank, "Contested Terrain," 106-7.

79 *Annual Report of the Minister of Mines,* 1897, 630.

80 Ibid., 1896, 594.

81 Dix, *Work Relations,* 102-3.

82 Patricia Roy, *A White Man's Province: British Columbia Politicians and Chinese and Japanese Immigrants, 1858-1914* (Vancouver: UBC Press, 1989), 78.

83 Ross Lambertson, "After *Union Colliery:* Law, Race, and Class in the Coalmines of British Columbia," in *Essays in the History of Canadian Law,* vol. 6, ed. Hamar Foster and John McLaren (Toronto: University of Toronto Press, 1995), 389.

84 Belshaw, "The British Collier," 28.

85 See *Annual Report of the Minister of Mines,* 1903 to 1905. In 1905, the rate improved to 4.8 for the Chinese.

86 *Report of the Royal Commission on Chinese and Japanese Immigration,* 76. Little does not indicate where the No. 2 mine was located, but it was at Cumberland. Belshaw refers to this all-Chinese mine in two of his articles, but incorrectly in both. In "The British Collier," 27, he refers to the "all-Chinese South Wellington" pit as having "one of the best accident records of any Island mine in the 1890s." Likewise, he states in "Mining Technique and Social Division on Vancouver Island, 1848-1900," *British Journal of Canadian Studies* 1 (1986), 56, that the Dunsmuirs "experimented with Chinese hewers at No. 2 Pit Wellington," when the source he cites (*Report of the Royal Commission on Chinese and Japanese Immigration,* 76) refers to Extension, and the "experiment" was the use of the Chinese as tracklayers, pushers "and that class of labour generally." Bowen, *Three Dollar Dreams,* 311, refers correctly to the Union mine.

87 Cited in Bowen, *Three Dollar Dreams,* 311.

88 *Annual Report of the Minister of Mines,* 1910, K218.

89 Ibid., K199.

90 Ibid., K214.

91 Ibid., K206.

92 *Nanaimo Free Press,* 6 October 1909.

93 *Annual Report of the Minister of Mines,* 1910, K194.

94 Ibid., K205-6. They argued that Thomas suffered steam burns (he was not singed) resulting from the humid condition of the mine. See also the extensive coverage in the *Nanaimo Free Press,* 23 October 1909.

95 *Annual Report of the Minister of Mines,* 1910, K219-20.

96 Ibid., K219. He also believed that some miners used coal and highly explosive coal dust as tamping material, instead of broken shale and clay, "the moment the Inspector's back is turned."

97 *Nanaimo Free Press,* 2 November 1909.

98 *Annual Report of the Minister of Mines,* 1910, K198 and K192. Ashworth made a number of recommendations, including improved ventilation and the compulsory use of safety lamps. However, he did not recommend increasing the number of inspectors, because he feared lowering the status of the inspector.

99 *Nanaimo Free Press,* 6 October 1909.

100 *Ladysmith Chronicle,* 10 October 1908, and *Annual Report of the Minister of Mines,* 1909, J240.

101 *Ladysmith Chronicle,* 9 October 1909.

102 Through the Workmen's Compensation Act, $1,500 was paid by the company per man. The Wellington-Extension Accident and Burial Fund paid $300 per man. Other monies raised included $500 from individuals in Ladysmith, $200 from the local branch of the Canadian Imperial Bank of Commerce, $700 from the City of Vancouver, $500 from the City of Victoria, $250 from the City of New Westminster, $500 from the Town of Ladysmith, and $140 from the Metropolitan Church in Victoria: *Ladysmith Chronicle,* 9 October 1909 and 13 October 1909.

103 Ibid., 6 October 1909.

104 See, for instance, John Benson, *British Coalminers in the Nineteenth Century* (London: Gill and Macmillan, 1980), 1-4; Long, *Where the Sun Never Shines,* 28; Ian McKay, "The Realm of Uncertainty: The Experience of Work in the Cumberland Coal Mines, 1873-1927," *Acadiensis* 16 (1986): 3.

105 McKay, "The Realm of Uncertainty," 40.

106 Bill Johnstone, *Coal Dust in My Blood: The Autobiography of a Coal Miner* (Victoria: Royal British Columbia Museum, 1993), 24.

107 As one German miner commented in 1912: "The miner's activity has this beauty to it: it is not routine work and with every action the intelligence can come into play. There are constantly occasions, in drilling, placing timbers, clearing falls, etc., where a good idea can save many hours work. Such activity offers satisfaction," quoted in Hickey, *Workers in Imperial Germany,* 116. Hickey continues: "He might have added that, unlike the factory worker, the miner was largely his own master when it came to the detailed planning and execution of his work; instead of being subject to the minute-by-minute supervision of the factory foreman, the hewer was one in a semi-autonomous group of colleagues. This combination of knowledge ... physical dexterity and skill, and reliance upon independent judgement could give the experienced miner a real sense of pride in his abilities and profession." See also Frank, "Contested Terrain," 106f.; McKay, "The Realm of Uncertainty," 40.

**Chapter 5: From Pillar to Post**

1 *Report of the Royal Commission on Industrial Disputes in the Province of British Columbia* (Ottawa: Department of Labour, 1903), Minutes of Evidence, 380. Hereafter cited as *Royal Commission, 1903.*

2 See, for instance, David J. Bercuson, "Labour Radicalism and the Western Industrial Frontier, 1897-1919," *Canadian Historical Review* 58 (1976): 253-54.

3 Allen Seager, "Socialists and Workers: The Western Canadian Coal Miners, 1900-1921," *Labour/Le Travail* 16 (1985): 55.

4 Bryan D. Palmer, *Working Class Experience: Rethinking the History of Canadian Labour, 1800-1991* (Toronto: McClelland and Stewart, 1992), 170. For the Maritimes, see Ian McKay, "Strikes in the Maritimes, 1901-1914," *Acadiensis* 13 (1983): 3-46.

5 James R. Conley, "Frontier Labourers, Crafts in Crisis and the Western Labour Revolt: The Case of Vancouver, 1900-1919," *Labour/Le Travail* 23 (1989): 28.

6 For example, Alan J. Wargo, "The Great Coal Strike: The Vancouver Island Coal Miners' Strike, 1912-14," BA thesis, University of British Columbia, 1962, 82.

7 As McKay claims: "A 19th century coal strike was a nuisance; a large coal strike in the 20th century was a calamity": McKay, "Strikes in the Maritimes," 10.

8 Stephen Hickey, "The Shaping of the German Labour Movement: Miners in the Ruhr," in *Society and Politics in Wilhelmine Germany,* ed. Richard J. Evans (London: Croom Helm, 1978), 227. For Britain, see Roy Church, "Edwardian Labour Unrest and Coalfield

Militancy, 1890-1914," *Historical Journal* 30 (1987): 841-57; Dick Geary, *European Labour Protest, 1848-1939* (London: Methuen, 1981), 106.

9 McKay, "Strikes in the Maritimes," 10 and 17.

10 John Belshaw, "The British Collier in British Columbia: Another Archetype Reconsidered," *Labour/Le Travail* 34 (1994): 17.

11 Lynne Bowen, *Three Dollar Dreams* (Lantzville, BC: Oolichan Books, 1987), 15-55.

12 Paul Philips, *No Power Greater: A Century of Labour in British Columbia* (Vancouver: Boag Foundation, 1967), 5; D.T. Gallacher, "Men, Money, Machines: Studies Comparing Colliery Operations and Factors of Production in British Columbia's Coal Industry to 1891," PhD thesis, University of British Columbia, 1979, 211. The VCMLC caused a public uproar when it hired Chinese strikebreakers, and was subsequently forced to settle with the miners.

13 In 1874, for instance, both the VCMLC and Dunsmuir employed only 489 men, 181 of whom were Chinese, and produced only 81,000 tons of coal.

14 See especially Bowen, *Three Dollar Dreams,* 157-58 and 233-34.

15 Belshaw, "The British Collier," 18.

16 This discussion of the Knights of Labor is heavily indebted to Bryan Palmer, *Working-Class Experience,* 127.

17 Ibid.

18 Belshaw, "The British Collier," 18.

19 John Douglas Belshaw, "British Coalminers on Vancouver Island, 1848-1900: A Social History," PhD thesis, University of London, 1987, 327-28.

20 Palmer, *Working-Class Experience,* 152.

21 In the early 1890s there was still a strong connection between the MMLPA and the Knights of Labor. See *Royal Commission, 1903,* 295. Belshaw, "The British Collier," 18, points out that "local assemblies continued to enjoy modest popularity among the miners into the 20th century."

22 Belshaw, "British Coalminers on Vancouver Island," 328.

23 See Jeremy Mouat, "The Politics of Coal: A Study of the Wellington Miners' Strike of 1890-91," *BC Studies* 77 (1988): 3-29; Bowen, *Three Dollar Dreams,* 331-48; Allen Seager and Adele Perry, "Mining the Connections: Class, Ethnicity, and Gender in Nanaimo, British Columbia, 1891," *Histoire sociale/Social History* 30 (1997). For instance, when the MMLPA tried to organize in Cumberland, five unionists were fired. See *Royal Commission, 1903,* 352.

24 See *Royal Commission, 1903,* 295. See also Belshaw, "The British Collier," 19, and Bowen, *Three Dollar Dreams,* 366. Lodges were disbanded in Wellington and East Wellington. Only the Northfield and Nanaimo locals survived.

25 Seager and Perry, "Mining the Connections," 63; Bowen, *Three Dollar Dreams,* 286.

26 *Nanaimo Free Press,* 14 April 1891, cited in Seager and Perry, "Mining the Connections," 63.

27 *Royal Commission, 1903,* 59.

28 Belshaw, "The British Collier," 19.

29 Although this was not without severe criticism by some union members, resulting in attacks on union leader Tully Boyce, eventually causing him to step down. See Bowen, *Three Dollar Dreams,* 367.

30 Mouat, "The Politics of Coal," 4; see also Jeremy Mouat, "The Genesis of Western Exceptionalism: British Columbia's Hard-Rock Miners, 1895-1903," *Canadian Historical Review* 71 (1990); and Jeremy Mouat, *Roaring Days: Rossland's Mines and the History of British Columbia* (Vancouver: UBC Press, 1995).

31 Belshaw especially emphasizes how the "labour aristocracy" and "penny capitalism" mitigated class conflict; see "British Coalminers on Vancouver Island." The notion of "labour aristocracy" builds on the work of Eric Hobsbawm and emphasizes the division between the well-paid skilled worker and the poorly paid unskilled worker. According to Belshaw, "The effect of these differentiations within the working class is to distance the labour aristocracy from the rest of the proletariat, to encourage 'embourgeoisement' (collaboration with and assimilation into the middle classes), and to broaden the appeal of 'respectability' among workers generally, all in a manner which is likely to weaken working-class unity and resolve in the class struggle": "British Coalminers on Vancouver Island," 334-35.

32 Belshaw, "The British Collier," 22.

33 McKenzie was a farmers' candidate, while Forster and Keith were miners. All three were supported by the MMLPA.

34 See Thomas Robert Loosmore, "The British Columbia Labor Movement and Political Action, 1879-1906," MA thesis, University of British Columbia, 1954, iii-vii, and Wargo, "The Great Coal Strike," 28-29.

35 The Reform Club was an association of pro-union groups in the district, including the miners' union, other local Nanaimo unions, and temperance societies. See Wargo, "The Great Coal Strike," 34-35.

36 Seager and Perry, "Mining the Connections," 68.

37 Palmer, *Working-Class Experience,* 124. Of course, racism was not monopolized by the miners.

38 *Nanaimo Free Press,* 8 April 1903.

39 See especially *Royal Commission, 1903.*

40 R.A.J. McDonald, *Making Vancouver, 1863-1913* (Vancouver: UBC Press, 1996), 113.

41 The last two years had been especially deadly for British Columbia's coal miners. In 1901 an explosion at Cumberland had taken the lives of 64 men, and in 1902, 125 men were killed in a horrific explosion at the Coal Creek colliery at Fernie in the Kootenays.

42 *Royal Commission, 1903,* 76 and 67.

43 *Nanaimo Free Press,* 9 April 1903.

44 Stuart M. Jamieson, *Times of Trouble: Labour Unrest and Industrial Conflict in Canada, 1900-66* (Ottawa: Task Force on Labour Relations, 1968), 115; Wargo, "The Great Coal Strike," 55.

45 *Nanaimo Free Press,* 10 February 1903 and 13 February 1903.

46 See *Royal Commission, 1903,* 343-47. Because negotiations between the men and the company over the agreement had run into difficulties, it was lying in abeyance at the time of Russell's testimony. For a copy of the proposed agreement, see *Royal Commission, 1903,* Exhibit 10, 753.

47 See, for example, the *Victoria Daily Times,* 26 February 1903, which reports that the miners held Baker in high esteem "for the manner in which he conducted the negotiations."

48 His refusal to meet with the committee led to this well-known and often quoted response by Dunsmuir: "I asked them where the nigger was. They said he had not come this time. I said I had heard there were objections going around that I would not see the deputation because there was a nigger on it. I told them I did not care whether it was composed of niggers, Chinese, Japanese, Indians or whitemen – I would see them as long as they were my own men": *Royal Commission, 1903,* 239.

49 *Nanaimo Free Press,* 28 April 1903; *Royal Commission, 1903,* 49.

50 *Royal Commission, 1903,* 481.

51 Nanaimo Community Archives, Western Federation of Miners, Enterprise Union, Minutes, 1903 [Mining, C3-B1].

52 *Royal Commission, 1903,* 57.

53 Ibid., 57-58.

54 Ibid., 481.

55 Ibid., 49, and Exhibit 13, 755.

56 *Victoria Daily Times,* 11 March 1903.

57 *Nanaimo Free Press,* 5 May 1903 and 6 May 1903. The newspaper reported that only two white miners were working and that white miners who had not joined the union were also leaving Cumberland. For the number of Chinese workers, see *Royal Commission, 1903,* 482. When Cumberland's manager, John Matthews, asked the Japanese why they were not working, they reportedly claimed that they were afraid the strikers would burn down "Jap Town" if they worked: *Royal Commission, 1903,* 482.

58 According to Matthews, almost 200 Chinese and Japanese had miners' certificates: *Royal Commission, 1903,* 501. On 15 July 1903, an explosion occurred in No. 6 shaft at Cumberland, killing sixteen Chinese miners and burning four others. Although the district was mined exclusively with locked safety lamps, "matches, tobacco and cigarettes were found in the working places and on the Chinamen that were killed; also an open safety lamp and one broken safety lamp in the place where the explosion occurred, showing negligence on their part": *Annual Report of the Minister of Mines of British Columbia,* 1904, H237. According to Carlos Schwantes, *Radical Heritage, Labor, Socialism and Reform in Washington and British Columbia, 1885-1917* (Seattle: University of Washington Press, 1979), 147, the manager was fined $25 for "illegally giving the Chinese mining permits."

59 *Victoria Daily Times,* 7 April 1903.

60 Nanaimo Community Archives, Western Federation of Miners. See also *Royal Commission, 1903,* 61-62; *Victoria Daily Times,* 19 May 1903.

61 *Victoria Daily Times,* 21 May 1903.

62 Ibid., 3 July 1903.

63 Ibid., 21 May 1903 and 22 May 1903.

64 Ibid., 3 July 1903.

65 For a more in-depth analysis of early working-class politics in British Columbia, see especially A. Ross McCormack, *Reformers, Rebels, and Revolutionaries: The Western Canadian Radical Movement, 1899-1919* (Toronto: University of Toronto Press, 1977), 18-34.

66 In his study of German social democracy, Dieter Groh describes this position within the dominant moderate right wing of the Social Democratic Party (SPD) of Germany, the largest socialist movement in the world at this time, as "revolutionary attentism," i.e., that socialists were content to wait for the inevitable collapse of capitalism. This followed the popular socialist theorist Karl Kautsky, who saw the SPD as a "revolutionary but not a revolution-making party." See Groh, *Negative Integration und revolutionärer Attentismus: Die deutsche Sozialdemokratie am Vorabend des 1. Weltkrieges* (Frankfurt am Main: Propyläen, 1973).

67 *Nanaimo Free Press,* 16 September 1903.

68 McCormack, *Reformers, Rebels, and Revolutionaries,* 63.

69 Hawthornthwaite does not appear to have been influenced by the revisionism of the German Social Democrat Eduard Bernstein (1850-1932), which, according to one scholar, rejected Marx's dialectical materialism and was based on the fact that social democracy was growing "not in an atmosphere of increasing misery and unemployment," as Marx had predicted, "but in one of unprecedented prosperity ... The enemy

of the working class was then not capitalism itself, not the capitalist state, but the small group of private interests which stubbornly refused to see the light of reason and social justice." See Carl E. Schorske's *German Social Democracy, 1905-1917: The Development of the Great Schism* (Cambridge, MA: Harvard University Press, 1955), 16 and 18.

70 *Nanaimo Free Press,* 31 August 1903.

71 Ibid., 16 September 1903.

72 Ibid., 31 August 1903.

73 Ibid., 16 September 1903.

74 Ibid., 31 August 1903.

75 Seager, "Socialists and Workers," 37.

76 On Hawthornthwaite and the VCMLC, see Loosmore, "The British Columbia Labor Movement," 176. More generally, see the recent account in Mark Leier, *Rebel Life: The Life and Times of Robert Gosden, Revolutionary, Mystic, Labour Spy* (Vancouver: New Star Books, 1999), 58-61 (sidebar).

77 *Nanaimo Free Press,* 31 August 1903.

78 *Vancouver Sun,* 8 February 1913.

79 Ibid.

80 *Ladysmith Chronicle,* 29 January 1910.

81 Belshaw, "British Coalminers on Vancouver Island," 342 and 346. The term "spittoon philosophers" was used by the communist activist George Hardy. See Leier, *Rebel Life,* 39 (sidebar).

82 *Royal Commission, 1903,* Testimony of Arthur Spencer, 348; Testimony of David Halliday, 412.

83 *Royal Commission, 1903,* Testimony of Samuel Mottishaw, 53.

84 Belshaw, "British Coalminers on Vancouver Island," 350.

85 Ibid., 351. According to Senator Templeman, Dunsmuir had no politics and "no party predilection, excepting for the party that will do most for him": quoted in Margaret Ormsby, *British Columbia: A History* (Toronto: Macmillan, 1958), 324.

86 On developments in Canada, see Palmer, *Working-Class Experience,* and Michael A. Cross and Gregory S. Kealey, eds., *The Consolidation of Capitalism, 1896-1929* (Toronto: McClelland and Stewart, 1983). For a concise summary, see also Craig Heron and Robert Storey, "On the Job in Canada," in *On the Job: Confronting the Labour Process in Canada,* ed. Craig Heron and Robert Storey (Montreal and Kingston: McGill-Queen's University Press, 1986), 12-18.

87 Palmer writes that "between 1900 and 1920, almost 200 consolidations absorbed approximately 440 firms": Palmer, *Working-Class Experience,* 158. See also Donald Reid, *The Miners of Decazeville* (Cambridge, MA: Harvard University Press, 1985), which discusses the impact of the "second industrial revolution" on French miners in great depth.

88 See Mark Leier, *Where the Fraser River Flows: The Industrial Workers of the World in British Columbia* (Vancouver: New Star Books, 1990), 13-15, for an excellent discussion of the impact of Taylorism, and, more recently, Cynthia Comacchio, "Mechanomorphosis: Science, Management, and 'Human Machinery' in Industrial Canada, 1900-45," *Labour/Le Travail* 41 (1998): 35-67.

89 By proletarianization, I do not mean the process of becoming a proletarian; rather, I am following Ian McKay, who defines it as the loss of power in the economic sphere. See Ian McKay, "Coal Miners and the Longue Durée: Learning from Decazeville," *Labour/Le Travail* 20 (1987): 223.

90 McKay, "Strikes in the Maritimes," 5.

91 Leopold H. Haimson and Charles Tilly, eds., *Strikes, Wars, and Revolutions in an International Perspective* (Cambridge: Cambridge University Press, 1989), 525-26. On coal miners' attempts to control the workplace, see David Frank, "Contested Terrain: Workers' Control in the Cape Breton Coal Mines in the 1920s," in Heron and Storey, *On the Job*, 102-23.

92 Andrew Yarmie, "The Right to Manage: Vancouver Employers' Associations, 1900-1923," *BC Studies* 91 (1991): 40-74; Andrew Yarmie, "The State and Employers' Associations in British Columbia: 1900-1932," *Labour/Le Travail* 45 (2000): 53-101; and McDonald, *Making Vancouver*. This point is also made in Haimson and Tilly, *Strikes, Wars, and Revolutions*, 527.

93 Yarmie, "The Right to Manage," 55.

94 R.A.J. McDonald, "Working Class Vancouver, 1886-1914: Urbanism and Class in British Columbia," *BC Studies* 69-70 (1986): 52-53; McDonald, *Making Vancouver*, 113-14. In 1903 James Dunsmuir agreed that unions interfered with the management of his business: "There is always a committee appointed to interfere with the management of the work. It is called a pit committee. They come around and say the men should have this, they should have that. They simply take the management out of the mine": *Royal Commission, 1903*, Minutes of Evidence, 240.

95 On the role of the state in capitalist societies, see Nicos Poulantzas, *Political Power and Social Classes* (London: Sheed and Ward, 1973). On the establishment of a liberal-capitalist legal structure and social order in colonial British Columbia, see Tina Loo, *Making Law, Order, and Authority in British Columbia, 1821-1871* (Toronto: University of Toronto Press, 1994).

96 Yarmie, "The State and Employers' Associations in British Columbia," 100.

97 Ormsby, *British Columbia: A History*, 324.

98 See especially the essay by John Belshaw, "Provincial Politics, 1871-1916," in *The Pacific Province*, ed. Hugh J.M. Johnston (Vancouver: Douglas and McIntyre, 1996), 156.

99 As pointed out by Yarmie, "The State and Employers' Associations," 73.

100 Dunsmuir's response was typical. He treated his employees like children. Inherently passive and easily led astray, his miners needed to be guided by the firm hand of their natural superiors because they were incapable of forming rational opinions for themselves. He agreed that they were being exploited, victimized, and manipulated, but not by ruthless capitalists and their agents, but by scurrilous foreigners, troublemakers, and union agitators: *Royal Commission, 1903*, Testimony of James Dunsmuir, 381.

**Chapter 6: The Great Strike**

1 On control strikes, see Ian McKay, "Strikes in the Maritimes, 1901-1914," *Acadiensis* 13 (1983): 18, 29-31. For a discussion of the different forms of control strike, see David Frank, "Contested Terrain: Workers' Control in the Cape Breton Coal Mines in the 1920s," in *On the Job: Confronting the Labour Process in Canada*, ed. Craig Heron and Robert Storey (Montreal and Kingston: McGill-Queen's University Press, 1986), 108ff.

2 John Norris, "The Vancouver Island Coal Miners, 1912-1914: A Study of an Organizational Strike," *BC Studies* 45 (1980): 56.

3 See BC Archives, GR-1323, B2101, 7529-16-13, File 34-35.

4 Alan J. Wargo, "The Great Coal Strike: The Vancouver Island Coal Miners' Strike, 1912-14," BA thesis, University of British Columbia, 1962, 71; *Report of Royal Commissioner on Coal Mining Disputes on Vancouver Island* (Ottawa: Government Printing Bureau, 1913), 16.

5 Letter to T.W. Crothers, Minister of Labour, 3 October 1912, in *Report of Royal Commissioner,* 13.

6 *Report of Royal Commissioner,* 16.

7 Letter to T.W. Crothers, Minister of Labour, 3 October 1912, in *Report of Royal Commissioner,* 11.

8 Management and the commission investigating the strike argued that the notice to remove tools was not a lockout. Rather it was "the usual notice for preservation of the tools and to save the men from being charged with the value thereof," and that it was "not at all in the nature of a voluntary dismissal of the men by the company": *Report of Royal Commissioner,* 15 and 17. If this had been the case, then the company would not have paid the men off "just as soon as the payroll could be made out." On the two-year contract, see Rev. John Hedley, *The Labor Trouble in Nanaimo District: An Address Given before the Brotherhood of Haliburton Street Methodist Church* (Nanaimo: n.p. 1913), 7.

9 *Report of Royal Commissioner,* 13. Foster refers to Clinton as the "czar of Cumberland" in *United Mine Workers Journal,* 31 October 1912.

10 There was some confusion about the duration of the protest holiday. According to Foster's testimony, "the men made one little mistake; they didn't say whether the holiday was one or two days": *Report of Royal Commissioner,* 14. The decision to declare a "holiday" rather than a strike was significant. Bowen suggests that the difference was merely semantic, that the miners' intention was clearly to strike, but in order to avoid being charged with declaring an illegal strike under the Industrial Disputes Investigation Act (IDIA), the term "holiday" was used: Lynne Bowen, *Boss Whistle* (Lantzville, BC: Oolichan Books, 1982), 137.

11 Letter from union leadership to Premier and Minister of Mines Richard McBride, 21 September 1912, in *Report of Royal Commissioner,* 13.

12 Wargo, "The Great Coal Strike," 68; *Report of Royal Commissioner,* 17.

13 *BC Federationist,* 5 October 1912. On Crothers's visit to Nanaimo, see *BC Federationist,* 13 July 1912. The minister of labour had been taken on a "tour" of No. 1 mine at Nanaimo in order to see the "local color."

14 Ibid., 5 October 1912.

15 Wargo, "The Great Coal Strike," 77.

16 McKay, "Strikes in the Maritimes," 41.

17 *Report of Royal Commissioner,* 24.

18 Wargo, "The Great Coal Strike," 83.

19 BC Archives, GR-1323, B2101, 7529-16-13, File 35.

20 Ibid.

21 *BC Federationist,* 28 September 1912.

22 On McBride's relationship with Mackenzie and Mann, see Patricia E. Roy, "Progress, Prosperity and Politics: The Railway Policies of Richard McBride," *BC Studies* 47 (1980): 3-28. See also R.B. Fleming, *The Railway King of Canada: Sir William Mackenzie, 1849-1923* (Vancouver: UBC Press, 1991).

23 Lynne Bowen, "The Great Vancouver Island Coal Miners' Strike, 1912-1914," *Journal of the West* 23 (1984): 34.

24 *Vancouver Sun,* 8 February 1913.

25 Ibid. See also *Report of the Royal Commission re: Coal in British Columbia* (Victoria: King's Printer, 1914).

26 See also Wargo, "The Great Coal Strike," 147-48.

27 See Priscilla Long, *Where the Sun Never Shines: A History of America's Bloody Coal Industry* (New York: Paragon House, 1989), 151-65; Maier B. Fox, *United We Stand: The United*

*Mine Workers of America, 1890-1990* (Washington, DC: United Mine Workers of America, 1990); and John H.M. Laslett, "British Immigrant Colliers, and the Origins and Early Development of the UMWA, 1870-1912," in *The United Mine Workers of America: A Model of Industrial Solidarity,* ed. John H.M. Laslett (University Park, PA: Pennsylvania State University Press, 1996), 29-50.

28 See Bruce Ramsey, *The Noble Cause: The Story of the United Mine Workers of America in Western Canada* (Calgary: UMWA District 18, 1990), 28-29; Allen Seager, "Socialists and Workers: The Western Canadian Coal Miners, 1900-1921," *Labour/Le Travail* 16 (1985): 27.

29 In 1908 the UMWA successfully challenged the existing Nova Scotia coal miners' union, the Provincial Workmen's Association (PWA). The first charter was issued to a Springhill local in December 1908, and in March 1909 District 26 of the UMWA was formed; over the course of the year, 4,500 miners had joined its ranks: *United Mine Workers Journal,* 20 January 1910; see also David Frank, "Industrial Democracy and Industrial Legality: The UMWA in Nova Scotia, 1908-1927," in Laslett, *The United Mine Workers of America,* 438-55. In May 1908 a referendum was held among PWA workers to decide whether to affiliate with the UMWA. The result was a 400-vote majority in favour of affiliation. The PWA leadership refused to recognize the outcome, however.

30 Ramsey, *The Noble Cause,* 36-37.

31 *BC Federationist,* 6 January 1912. See also *Report of Royal Commissioner,* 17.

32 T.L. Lewis's stint as union president came to an ignominious end in January 1911. The union was heavily in debt and Lewis was accused of working for the operators, which he vehemently denied. After his failed re-election bid, he joined the West Virginia Coal Operators' Association. See Long, *Where the Sun Never Shines,* 259-61.

33 *BC Federationist,* 6 January 1912; *United Mine Workers Journal,* 18 January 1912; see also Hedley, *The Labor Trouble in Nanaimo District,* 5. Not much is known about Pettigrew. He is described as an "international organizer" by Bowen, but is listed in the provincial voters' list for 1911 as a miner living on First Avenue, Ladysmith.

34 *Report of Royal Commissioner,* 17.

35 Working conditions and safety remained burning issues. Miners complained of inadequate enforcement of existing safety regulations, unacceptable levels of dust and water in some mines, and the lack of proper sanitation facilities underground and showering facilities above ground. In addition, they worried about the extremely high cost of powder and the lack of proper compensation to accident victims and their families. See, for instance, *BC Federationist,* 13 July 1912 and 3 August 1912; *United Mine Workers Journal,* 20 June 1912; see also *BC Federationist,* 20 July 1912 and 17 August 1912.

36 Wargo, "The Great Coal Strike," 62.

37 *BC Federationist,* 3 August 1912.

38 Ibid., 17 August 1912 and 14 September 1912.

39 Wargo, "The Great Coal Strike," 81-82; according to Norris, "The Vancouver Island Coal Miners," 65, the international, "though supporting the action of the local in public, privately deplored it."

40 Wargo, "The Great Coal Strike," 94.

41 Quoted in Wargo, "The Great Coal Strike," 98.

42 BC Archives, GR-1323, B2087, 9318-16-12, File 44.

43 Wargo, "The Great Coal Strike," 81; Norris, "The Vancouver Island Coal Miners," 57.

44 Bowen, *Boss Whistle,* 137-38.

45 BC Archives, A.F. Buckham Personal and Professional Papers, Box 21, File 3: W.L. Coulson to J.R. Lockhart, 30 September 1912; J.R. Lockhart to W.L. Coulson, 1 October

1912. Lockhart listed among the top agitators Foster, Pettigrew, Irvine, Patterson, and Angelo, all of whom were senior union officials, as well as seven miners, two fire bosses, a hoist engineer, and a checkweighman.

46 Wargo, "The Great Coal Strike," 88.
47 *United Mine Workers Journal,* 3 October 1912.
48 Ibid., 14 November 1912.
49 *BC Federationist,* 5 October 1912; see also Bowen, *Boss Whistle,* 138, and Wargo, "The Great Coal Strike," 86.
50 Long, *Where the Sun Never Shines,* 250-51.
51 "We extend to all men in and around the mines without regard to race or color, except Chinese and Japanese, an invitation to unite with us that these ends may be attained," cited in Ramsey, *The Noble Cause,* 30. It appears that the Lethbridge local of District 18 admitted Chinese workers in 1906 during the Lethbridge strike: ibid., 42. See also Patricia Roy, *A White Man's Province: British Columbia Politicians and Chinese and Japanese Immigrants, 1858-1914* (Vancouver: UBC Press, 1989), 253-54.
52 *BC Federationist,* 20 July 1912.
53 Ibid., 17 August 1912.
54 Ibid., 2 November 1912.
55 Ibid., 21 September 1912; *United Mine Workers Journal,* 3 October 1912 and 31 October 1912.
56 Letter and telegram in *BC Federationist,* 9 November 1912.
57 Ibid., 1 November 1912.
58 *United Mine Workers Journal,* 21 November 1912 and 16 January 1913.
59 See, for example, David Gilbert, *Class, Community, and Collective Action* (Oxford: Oxford University Press, 1992).
60 *Vancouver Sun,* 12 February 1913.
61 *Report of the Royal Commission on Labour* (Victoria: King's Printer, 1914), M1.
62 *United Mine Workers Journal,* 16 January 1913.
63 *Report of the Royal Commission on Labour,* M18.
64 *United Mine Workers Journal,* 3 April 1913.
65 The report concluded that there was no evidence to suggest that members of the radical IWW were attempting "to carry into practice the principles of this organization": *Report of the Royal Commission on Labour,* M2-M3.
66 *Report of Royal Commissioner,* 21; Wargo, "The Great Coal Strike," 106-7.
67 *United Mine Workers Journal,* 29 May 1913; *Nanaimo Free Press,* 2 May 1913.
68 *Report of Royal Commissioner,* 23.
69 Ibid., 18 and 23.
70 *Nanaimo Free Press,* 2 May 1913.
71 Ibid., 3 May 1913.
72 Ibid., 9 May 1913.
73 Ibid., 12 May 1913; *Report of Royal Commissioner,* 19.
74 Wargo, "The Great Coal Strike," 111.
75 *Nanaimo Free Press,* 12 May 1913.
76 *BC Federationist,* 6 June 1913.
77 *Nanaimo Free Press,* 9 May 1913 and 12 May 1913.
78 *United Mine Workers Journal,* 19 June 1913.
79 Even though the Price Commission discounted the opinion "that the officials of the United Mine Workers' organization are in the pay of the mine owners of Washington State": *Report of Royal Commissioner,* 22.

80 *United Mine Workers Journal,* 27 March 1913.
81 *Nanaimo Free Press,* 12 May 1913.

**Chapter 7: No Ordinary Riot**

1 *Nanaimo Free Press,* 23 August 1913.
2 John Norris, "The Vancouver Island Coal Miners, 1912-1914: A Study of an Organizational Strike," *BC Studies* 45 (1980): 57 and 71.
3 Allen Seager, "Socialists and Workers: The Western Canadian Coal Miners, 1900-1921," *Labour/Le Travail* 16 (1985): 30.
4 *Nanaimo Free Press,* 12 May 1913 and 11 August 1913.
5 *Nanaimo Free Press,* 11 August 1913; quote from *Vancouver Sun,* 12 February 1913.
6 BC Archives, GR-1323, B2087, 9318-16-12, 10 and 17.
7 See Lynne Bowen, *Boss Whistle* (Lantzville, BC: Oolichan Books, 1982): 146-49.
8 Letter in *United Mine Workers Journal,* 31 July 1913.
9 *Nanaimo Free Press,* 28 July 1913; Alan J. Wargo, "The Great Coal Strike: The Vancouver Island Coal Miners' Strike, 1912-14," BA thesis, University of British Columbia, 1962, 112. See also the lengthy statement by Robert Foster in *Victoria Daily Times,* 21 August 1913.
10 According to an article in the *BC Federationist,* 3 October 1913: "It may be noted that this is the third or fourth profitable government commission which Mr. Crother's [sic] law partner has secured since the minister of labor assumed office. His work for the government will, it is understood, continue for some time yet, and will probably involve a trip to England this month."
11 *BC Federationist,* 12 September 1913.
12 *United Mine Workers Journal,* 11 September 1913. According to the paper, Clinton "urged the mounted police to ride down and shoot the strikers." Foster's claim is in the *Victoria Daily Times,* 21 August 1913.
13 *United Mine Workers Journal,* 11 September 1913; *BC Federationist,* 12 September 1913. See also Bowen, *Boss Whistle,* 152-53. Bowen states that Cave was a special policeman. The newspapers report that he was a strikebreaker.
14 BC Archives, GR-1323, B2087, 9318-16-12, Files 88 and 89. Police in Cumberland reported that they did not receive any assistance or support from local authorities, including the mayor and police commissioner, as these men were apparently "rank Socialists, and were elected on the Socialistic ticket." The police also reported that "a special Constable (name not known) was on duty on the bridge, but appears to have been absent for a few minutes at the time of the explosion."
15 *Western Clarion,* 27 September 1913; *BC Federationist,* 19 September 1913.
16 *Western Clarion,* 27 September 1913; *BC Federationist,* 19 September 1913. As Seager, "Socialists and Workers," 44, points out, "Only the socialist press deemed [Campbell's] appeals worthy of publication."
17 BC Archives, GR-0429, Box 19, File 1, 7529/16, Pinkerton Reports, 26 August 1913.
18 *United Mine Workers Journal,* 11 September 1913.
19 BC Archives, Pinkerton Reports, 26 August 1913, 27 August 1913, 12 September 1913.
20 See, for instance, ibid., 13 October 1913, where the agent reported that "the Police officers [in Ladysmith] were unduly careful not to give offense to defendants and I can repeat at this time that I have seen since many manifestations of weakness of the same sort on the part of the officers testifying. It seems strange that with so many crimes being committed and within their very sight that they can do no more than identify an occasional prisoner as 'being there' but 'doing nothing wrong'. I have

heard time and again that 'Hannay is a good guy' and meant to imply that he does not tell all that he knows regarding these men. Ike Portrey, himself admitted to me that constable Hannay 'did not go as strong' as he could in this respect."

21 BC Archives, GR-1323, B2087, 9318-16-12, File 91.

22 BC Archives, GR-445, Box 5, File 3, Provincial Police Reports, July 1913.

23 Nanaimo Community Archives, Oral History Project, Interview 84-6-8, Viola Johnson-Cull.

24 *BC Federationist,* 27 June 1913. In Cumberland, two women were tried for assault and found guilty. According to the Crown prosecutor, T.B. Shoebotham, "the trouble was what to do with them. Knowing of the Prison Congestion, I felt it to be useless to have them sentenced to imprisonment. To impose a fine meant that their sympathizers would pay it, and would have no restraining effect on that account. I therefore impressed upon the Magistrate the necessity of putting them on their good behavior and allowing them out on suspended sentence, which he did. It has had a salutary effect and there has been no trouble with women since." See BC Archives, GR-1323, B2087, 9318-16-12, File 20.

25 BC Archives, GR-1323, B2087, 9318-16-12, File 49.

26 *Nanaimo Free Press,* 6 August 1913 and 7 August 1913.

27 *BC Federationist,* 12 August 1913; *United Mine Workers Journal,* 21 August 1913. See also Bowen, *Boss Whistle,* 154.

28 *Nanaimo Free Press,* 11 August 1913.

29 Ibid., 12 August 1913; *Victoria Daily Times,* 13 August 1913.

30 *Victoria Daily Times,* 13 August 1913.

31 *Nanaimo Free Press,* 13 August 1913.

32 See Bowen, *Boss Whistle,* 157-59.

33 BC Archives, GR-1323, B2101, 7529-16-13, File 179.

34 Bowen, *Boss Whistle,* 160 and 182; *Vancouver Sun,* 1 September 1913.

35 *Victoria Daily Times,* 29 August 1913 and 30 August 1913.

36 See BC Archives, GR-0518, Box 5, File 111, 33, "Commission on Claims Arising Out of Riots in 1913 and 1914 on Vancouver Island," for the extensive list of his life-threatening injuries. He remained in hospital for many months and was eventually compensated $7,500 for his personal injuries and given a job with the fisheries department.

37 *Nanaimo Free Press,* 13 August 1913.

38 See BC Archives, Pinkerton Reports, 29 August 1913 and 1 September 1913.

39 Quoted in Bowen, *Boss Whistle,* 161; NCA, Interview with Viola Johnson-Cull.

40 BC Archives, GR-1323, B2101, 7529-16-13, File 187. See also Bowen, *Boss Whistle,* 161, and the story in the *Victoria Daily Colonist,* 5 February 1961.

41 *Victoria Daily Times,* 13 August 1913.

42 *Vancouver Sun,* 15 August 1913. See also evidence given at the preliminary hearings in Ladysmith in the *Victoria Daily Times,* 29 August 1913.

43 BC Archives, GR-1323, B2102, 7529-16-13, File 180.

44 Ibid. According to Leonard de Gex, manager of the Canadian Imperial Bank of Commerce in Ladysmith, the strikers submitted a resolution to a citizens' meeting, declaring: "Whereas, this city has been disturbed by industrial disturbances, and whereas further disturbance is unnecessary, providing the mayor and city officials will promise to use their influence towards keeping the militia away, the union will promise to keep order and not to molest property": *Victoria Daily Times,* 30 August 1913.

45 *Vancouver Sun,* 15 August 1913.
46 See Bowen, *Boss Whistle,* 164-66.
47 *Nanaimo Free Press,* 13 August 1913.
48 Baxter, who spent ten days in hospital, later claimed compensation. His claim was rejected, however: "There is evidence that he sympathized with the strikers, if he did not actually take part in their doings. He mixed freely with them, and one of them had suggested to him that he take a gun. He had no business at the time being within four miles of the disturbances. He had heard that there was shooting going on, and had actually been warned twice not to proceed": BC Archives, GR-0518, Box 5, File 111, 31, "Commission on Claims Arising Out of Riots"; *Victoria Daily Times,* 14 August 1913 and 21 August 1913. The union complained that the police did not even try to find out who shot him, but the *Victoria Daily Times,* 15 August 1913, reported that "a couple of men are under arrest on a more or less definite charge of having shot Alexander [sic] Baxter, the inquisitive Nanaimo contractor who looked down the Extension tunnel shaft and stopped the flight of a bullet."
49 The *Victoria Daily Times,* 15 August 1913, claimed that fourteen cottages had burned down and others had been badly damaged, and that the bunkhouses and the pitheads had also been destroyed.
50 *Vancouver Sun,* 15 August 1913. The following day, the *Sun* reported that all was quiet and that Colonel R.G. Leckie had claimed that "conditions had been greatly exaggerated." Nonetheless 300 additional militiamen were being dispatched: *Vancouver Sun,* 16 August 1913.
51 Bowen, *Boss Whistle,* 183; David Ricardo Williams, *Call in Pinkerton's: American Detectives at Work for Canada* (Toronto: Dundurn Press, 1998), 157.
52 *Vancouver Sun,* 22 August 1913. See also Parker Williams's more moderate comments: *Vancouver Sun,* 2 September 1913.
53 *Nanaimo Free Press,* 15 August 1913. In many respects the attorney general was an antidote to McBride, the consummate politician who prided himself on his wit and charm and who sought public acclaim. Bowser, in contrast, was a political realist, a man content to operate behind the scenes and for whom power in and of itself was the ultimate aphrodisiac. In the harsh yet colourful words of historian Martin Robin, Bowser was "a bulldog, a roughhouse fighter and street brawler who marshalled his considerable strength, intelligence and pugnacity to pummel his opponents into oblivion in the House and on the hustings. He was a frank and open political thug who thrived in the slough of patronage and pay-off which, to his mind, formed the pith and substance of politics." A man with "scant respect for legal traditions and the rule of law," Bowser was "a brutal fellow who preferred power to fame and left the limelight to more gracious persons," such as Richard McBride: Martin Robin, *The Rush for Spoils: The Company Province, 1871-1933* (Toronto: McClelland and Stewart, 1972), 132.
54 *Vancouver Sun,* 15 August 1913 and 18 August 1913. Other people compared Nanaimo under the military occupation to Czarist Russia: "One might as well be in Russia as in Nanaimo tonight, so rigid has the military discipline become": *Nanaimo Free Press,* 23 August 1913; *Vancouver Sun,* 18 August 1913.
55 *Victoria Daily Times,* 21 August 1913 and 23 August 1913. But that was not all. In September, complaints were heard about the deportment of the fifty Seaforth Highlanders who remained in Nanaimo. According to the spy employed by the attorney general, the "kilties," also known as the "half-clad barbarians," offended the sensibilities of Nanaimo's more respectable citizens because they wore "no covering over their upper

legs and hips other than the kilts. They wear a narrow strip of light colored cloth hung from the front of the belt to the rear and fitting tight over their privates. That is all. Now, their camp site is elevated some four to ten feet above the level of the street and on a slope opposite the post office. These fellows seem to forget these conditions and if engaged in tightening a guy rope on the tent or fixing a peg or lying around in a careless manner, or stoop over, their skirts go up so far behind that all passers by, women, children and the people seated opposite are given a display that one is not supposed to meet with in a civilized land": BC Archives, Pinkerton Reports (severed copy), 13 September 1913.

56 *Vancouver Sun,* 21 August 1913; *Nanaimo Free Press,* 23 August 1913.

57 *Vancouver Sun,* 29 August 1913 and 30 August 1913.

58 Ibid., 19 August 1913; Bowen, *Boss Whistle,* 177-78.

59 *Vancouver Sun,* 22 August 1913.

60 Ibid., 28 August 1913.

61 The best account remains Bowen, *Boss Whistle,* 181; *Vancouver Sun,* 22 August 1913.

62 Bowen, *Boss Whistle,* 162-64. Others suggested that Guthrie was the one who led the men around town. The friendly relationship between the Ladysmith policemen and the strikers was commented upon by the Pinkerton agent: BC Archives, Pinkerton Reports (severed copy), 13 October 1913.

63 See Bowen, *Boss Whistle,* 188-89.

64 Ibid., 189. When some Nanaimo strikers were acquitted in December 1913, they were met by "hundreds of marchers singing the Marseillaise, as they marched and men, women displaying red badges": BC Archives, Pinkerton Reports, 3 December 1913.

65 See, for instance, BC Archives, Pinkerton Reports, 9 September 1913.

66 *Vancouver Sun,* 25 August 1913 and 26 August 1913.

67 Ibid., 1 September 1913.

68 BC Archives, Pinkerton Reports, 23 August 1913, 24 August 1913, 29 August 1913, 5 September 1913; Wargo, "The Great Coal Strike," 135. Norris, "The Vancouver Island Coal Miners," 56, argues, in contrast, that the riots "further weakened the position of the strikers and the union and further encouraged the return to work of strikers, while, by isolating the remainder, strengthening their resolve to continue the strike long after it was lost." While I agree that the strikers' resolve was strengthened, I disagree that their position was fundamentally weakened by the riots and that they were isolated as strikers returned to work. Finally, I disagree that the strike was already lost. If this was the case, why did it last another full year?

69 BC Archives, GR-0068, vol. 2, Operative 29, 29 November 1913.

70 BC Archives, GR-1323, B2101, 7529-16-13, File 174, "Remarks of Judge Howay Relative to the Sentencing of Ladysmith Strikers."

71 Bowen, *Boss Whistle,* 187 and 163.

72 Norris, "The Vancouver Island Coal Miners," 71.

73 *Nanaimo Free Press,* 13 August 1913.

74 Norris, "The Vancouver Island Coal Miners," 69.

75 *Vancouver Sun,* 18 August 1913.

76 Norris, "The Vancouver Island Coal Miners," 68.

77 Mark Leier, *Rebel Life: The Life and Times of Robert Gosden, Revolutionary, Mystic, Labour Spy* (Vancouver: New Star Books, 1999), 30.

78 *BC Federationist,* 19 September 1913.

79 See Leier, *Rebel Life,* 38.

80 Ibid., 30-31.
81 *Vancouver Sun,* 16 August 1913.
82 BC Archives, Pinkerton Reports, 13 November 1913 and 14 November 1913.
83 Ibid., 21 October 1913.
84 *BC Federationist,* 22 August 1913.
85 BC Archives, Pinkerton Reports, 6 December 1913. Indeed, the UMWA denounced Gosden's "coffee speech," and in January 1914 he resigned from the presidency of the Miners' Liberation League. See Leier, *Rebel Life,* 32. The *Vancouver Sun,* 16 August 1913, reported that on the previous day, "several I.W.W. leaders were understood to have made an attempt to organize a mob from among themselves to go down to the wharves and ridicule the militia who had been called out. The efforts, however, fell rather flat, as there was but a small turnout to jeer at the troops."
86 *Vancouver Sun,* 14 August 1913.
87 Priscilla Long, "The 1913-1914 Colorado Fuel and Iron Strike, with Reflections on the Causes of Coal-Strike Violence," in *The United Mine Workers of America: A Model of Industrial Solidarity,* ed. John H.M. Laslett (University Park, PA: Pennsylvania State University Press, 1996), 364, writes that "violence was most emphatically not the policy of the union, district or national."
88 *United Mine Workers Journal,* 8 May 1913.
89 The issue of union violence is also addressed in William M. Baker, "The Miners and the Mounties: The Royal North West Mounted Police and the 1906 Lethbridge Strike," *Labour/Le Travail* 27 (1991): 75ff., which effectively demonstrates how union leader Frank Sherman took vigorous steps to maintain order during the Lethbridge strike of 1906 and that the UMWA opposed violence by its members.
90 *BC Federationist,* 3 October 1913.
91 Quoted in Bruce Ramsey, *The Noble Cause: The Story of the United Mine Workers of America in Western Canada* (Calgary: UMWA District 18, 1990), 86.
92 BC Archives, Pinkerton Reports, 9 September 1913.
93 Norris, "The Vancouver Island Coal Miners," 69.
94 BC Archives, Pinkerton Reports, 6 September 1913.
95 These arguments are by no means new. Klaus Tenfelde, for instance, describes the Herne Riots of 1899 in Germany as a form of social protest: Tenfelde, "The Herne Riots of 1899," in *The Social History of Politics: Critical Perspectives in West German Historical Writing since 1945,* ed. Georg G. Iggers (Leamington Spa, UK: Berg Publishers, 1985), 282-312. See also Heinrich Volkmann and Jürgen Bergmann, eds., *Sozialer Protest: Studien zu traditioneller Resistenz und kollektiver Gewalt in Deutschland* (Opladen: Westdeutscher Verlag, 1984); Dick Geary, "Protest and Strike: Recent Research on 'Collective Action' in England, Germany, and France," in *Arbeiter und Arbeiterbewegung im Vergleich,* ed. Klaus Tenfelde (Munich: Beck Verlag, 1986), 363-87. There is considerable debate about the use of state violence in Canadian labour disputes. According to a recent discussion, the state called out the militia in only 1 percent of strikes between 1895 and 1914, and although the Canadian militia were prepared to kill in defence of capitalism and the political status quo, only one striker was killed by them between 1867 and 1914. In contrast to that in other industrial nations, state violence in Canadian labour disputes was rare indeed. See Baker, "The Miners and the Mounties."
96 My discussion of crowd actions is informed by E.P. Thompson, "The Moral Economy of the English Crowd," in *Customs in Common* (New York: New Press, 1993), 238. There is a vast literature on crowd action and its symbolic content. See, most recently, Clark

McPhail, *The Myth of the Madding Crowd* (New York: A. de Gruyter, 1991); and Robert Rutherdale, "Canada's August Festival: Communitas, Liminality, and Social Memory," *Canadian Historical Review* 77 (1996): 221-49.

97 A point made by Rutherdale, "Canada's August Festival," 239, in his discussion of the enthusiastic crowds in August 1914. Accordingly, the participants experienced "liminality," that is, a sort of transcendent "time out of time." Liminality is often seen as a rite of passage, a carnivalesque "liberation" in which the existing social order is symbolically subverted and overturned.

98 The following discussion summarizes my article "'Stout Ladies and Amazons': Women in the British Columbia Coal-Mining Community of Ladysmith, 1912-14," *BC Studies* 114 (1997). On the Auxiliaries, see Sara Diamond, "A Union Man's Wife: The Ladies' Auxiliary Movement in the IWA, the Lake Cowichan Experience," in *Not Just Pin Money: Selected Essays on the History of Women's Work in British Columbia,* ed. Barbara K. Latham and Roberta J. Pazdro (Victoria: Camosun College, 1984), 287-96; and Stephane E. Booth, "Ladies in White: Female Activism in the Southern Illinois Coalfields, 1932-1938," in Laslett, *The United Mine Workers of America,* 371-92. These activities placed women in an awkward position within the community. The men acknowledged women's important contribution to the struggle, but feared their challenge to established gender roles. Men often viewed women as a threat to class solidarity, because of their stereotypical social aspirations, seductive power, and corruptibility. They were suspicious of "the ability of the women to communicate among themselves and bind together," and often derided these women, especially the wives of miners, as troublemakers, gossips, and pests. See Steven Penfold, "'Have You No Manhood In You?' Gender and Class in the Cape Breton Coal Towns, 1920-1926," *Acadiensis* 23 (1994): 29. This point is also made by Mark Leier, *Red Flags and Red Tape: The Making of a Labour Bureaucracy* (Toronto: University of Toronto Press, 1995), 140. Cynthia Bindocci cites an article from the *Coal Age* of 28 March 1918, which states: "One woman of this sort can cause more trouble and do more to impede the smooth running of a mining plant than would seem possible. And every community seems to have at least one of these pests"; see Bindocci, "A Comparison of the Roles of Women in Anthracite and Bituminous Mining in Pennsylvania, 1900-20," in *Sozialgeschichte des Bergbaus im 19. und 20. Jahrhundert,* ed. K. Tenfelde (Munich: Beck Verlag, 1992), 686. See also Star Rosenthal, "Union Maids: Organized Women Workers in Vancouver, 1900-1915," *BC Studies* 41 (1979): 37. As Freda Hogan, an American writing in the *United Mine Workers Journal,* 8 February 1912, put it: "No woman has ever felt the keenness of the class struggle more, or knows better that there is a class struggle than does the miner's wife."

99 *Vancouver Sun,* 1 September 1913.

100 Bowen, *Boss Whistle,* 168.

101 *Vancouver Sun,* 31 January 1914.

102 BC Archives, GR-1323, B2101, 7529-16-13, File 176. In his report to the federal minister of justice, Howay stated that he viewed Charles Axelson and his wife "as particularly dangerous people," File 180. In a controversial statement made to the *New Westminster News* at the end of the trial, he emphasized once again the role played by women: "Fully 90 per cent of the women ranked with the men in disregard for property and life": *New Westminster News,* 27 October 1913. These comments convinced the government to try the Nanaimo men in New Westminster and to replace Howay with Morrison.

103 As the *BC Federationist* claimed, in an attempt to drum up sympathy for its cause, the state and the companies were waging a "war on women": *BC Federationist,* 19 September 1913.

104 BC Archives, GR-0429, Box 19, Pinkerton Reports, 25 August 1913.

105 Ibid., 12 September 1913.

106 Ibid., 29 August 1913 and 3 September 1913. Women who appeared to transgress middle-class codes of female conduct were invariably referred to as "amazons." See also Bowen, *Three Dollar Dreams,* 165, which records a reporter's impression of an eviction during the 1877 strike: "No sooner had he [a deputy] made an entry, than he was pitchforked out of the house and kicked and cuffed through a line of indignant Amazons, when he quickly beat a retreat ... None of the parties were hurt, although they received considerable cuffing from the women who vigorously assaulted them."

107 *Vancouver Sun,* 25 August 1913.

108 The demand for the vote was not new. Equal suffrage had been an issue since British Columbia joined Confederation in 1871, and it was a concern of the working class in Ladysmith and Nanaimo. The legislature voted on female suffrage at least five times during the 1890s; twice in the new century, James Hawthornthwaite unsuccessfully introduced bills to enfranchise women. Equal suffrage was often an important issue for women in earlier coal strikes. During the 1891 strike, for example, a group of eighteen women, one of whom, Naomi Poulet, was assaulted by a young strikebreaker, led a "March for Female Suffrage" along the road from Nanaimo to Wellington; in 1894, votes for women was the first item on the agenda of the Workingman's Platform, a coalition of political opposition groups in Nanaimo, which declared: "That all women, resident within the Province and being British subjects of the full age of twenty-one years, be entitled to vote at the Provincial elections upon the same terms and conditions that men are so entitled": Michael H. Cramer, "Public and Political: Documents of the Women's Suffrage Campaign in British Columbia, 1871-1917: The View from Victoria," in *In Her Own Right: Selected Essays on Women's History in British Columbia,* ed. Barbara Latham and Cathy Kess (Victoria: Camosun College, 1980), 79; Irene Howard, *The Struggle for Social Justice in British Columbia: Helena Gutteridge, the Unknown Reformer* (Vancouver: UBC Press, 1992), 60. See Allen Seager and Adele Perry, "Mining the Connections: Class, Ethnicity, and Gender in Nanaimo, British Columbia, 1891," *Histoire sociale/Social History* 30 (1997): 58ff.; Thomas Robert Loosmore, "The British Columbia Labor Movement and Political Action, 1879-1906," MA thesis, University of British Columbia, 1954, xi.

109 Howard, *The Struggle for Social Justice,* 65.

110 *Vancouver Sun,* 12 February 1913.

111 Ibid.

112 Ibid.

113 Ibid.

114 Ibid.

115 When the union newspaper urged women to agitate for the cause, it was "for the protection of your home and your family": *United Mine Workers Journal,* 11 January 1912.

116 *Nanaimo Free Press,* 7 November 1913 and 10 November 1913. On Gosden's and the IWW's influence, see Leier, *Rebel Life,* 30-33, and A. Ross McCormack, *Reformers, Rebels, and Revolutionaries: The Western Canadian Radical Movement* (Toronto: University of Toronto Press, 1977), 114.

117 *Nanaimo Free Press,* 3 November 1913 and 5 November 1913. The union even offered to select a non-union member for the board, but this too was rejected.

118 According to union officials, attempts to entice strikebreakers to Vancouver Island failed for the most part. George Pettigrew and other union officials constantly discouraged

strikebreakers from coming to the Island. Those looking for work at local labour agencies were kept fully informed about the strike situation on the Island, and many returned to the mainland or sought other work. Estimates of the number of men who returned to work vary. See Norris, "The Vancouver Island Coal Miners," 60. On strikebreakers, see Bowen, *Boss Whistle,* 146-47. See also *Report of Royal Commissioner on Coal Mining Disputes on Vancouver Island* (Ottawa: Government Printing Bureau, 1913), 24, which was disingenuous in a number of ways. By emphasizing the number of Asians given certificates, for instance, Commissioner Samuel Price was seeking to deflect attention from the fact that Asians now constituted 62 percent of the workforce (as opposed to less than 20 percent prior to the strike). He also completely ignored the fact that the employment of Asians underground was illegal. For a criticism of the report, see *United Mine Workers Journal,* 25 December 1913. BC Archives, MS-0436, A.F. Buckham Personal and Professional Papers, Vol. 21, File 3, contains a contract between CC(D) and forty-six men from Alberta dated 17 February 1913.

119 Bowen, *Boss Whistle,* 190. Production was not completely normalized. According to an internal report, the Cumberland mines were "in shape to produce 2,500 tons per day," which was greater than the former capacity, but poor market conditions due to lighter shipping, a mild winter, and the economic recession prevented this. CC(D) therefore planned to shift men from Cumberland to Extension, where production had recently resumed after the repair of damage to the mine through falls and flooding caused by the lengthy idle time. By mid-March 1914, however, production was only 37 percent of the prestrike level, while at Nanaimo's Western Fuel, production was 50 percent. See BC Archives, MS-0436, A.F. Buckham Personal and Professional Papers, Vol. 33, File 11. See also *The Labour Gazette,* February 1914, 920-21; March 1914, 1079.

120 Referred to by the *Canadian Annual Review* of 1914 as "a female anarchistic orator from the States," Mother Jones, by now eighty-three years of age, had spent thirty-five years campaigning throughout the United States for union rights and better working conditions. In 1914 alone, she travelled more than seventy thousand miles, was imprisoned in Colorado, where she had been agitating in support of striking miners, and was granted an audience with US President Woodrow Wilson. See *Nanaimo Free Press,* 5 June 1914, 6 June 1914, and 8 June 1914; *Victoria Daily Times,* 6 June 1914 and 11 June 1914. See also Philip S. Foner, ed., *Mother Jones Speaks: Collected Writings and Speeches* (New York: Monad Press, 1983), 250-57.

121 *Victoria Daily Times,* 23 June 1914, which states: "Although having voted to continue the strike [by a majority of 1,250 votes] the strikers feel confident that word will soon be received from Indianapolis to the effect that relief money has been stopped, in which case the strike will at once collapse as the strikers will feel that they have been finally turned down by the international organization, and find it impossible to further continue the struggle without financial assistance."

122 *Nanaimo Free Press,* 21 July 1914.

123 *Victoria Daily Times,* 15 July 1914; *Nanaimo Free Press,* 20 August 1914.

**Conclusion**

1 Isabelle Davis, "Forty-Ninth Parallel City: An Economic History of Ladysmith," BA thesis, University of British Columbia, 1953, 7.

2 Richard Goodacre, *Dunsmuir's Dream* (Victoria: Porcépic Books, 1991), 76.

3 The Morden Colliery, situated to the east of South Wellington, opened during the strike but closed down in 1922. Granby mine, in Cassidy, opened in 1918 and lasted

until 1932. The Reserve Mine in Chase River was operated by Western Fuel and began production during the strike. It was closed for four years during the Depression, but did not survive a 1939 fire. See especially Lynne Bowen, *Boss Whistle* (Lantzville, BC: Oolichan Books, 1982), 199-232.

4 By the early 1920s, fuel oil was being extensively used in BC. This prompted one author to write in 1922: "The introduction of fuel-oil on the Pacific coast has probably done more to retard the development of the Vancouver Island coal industry than anything else and accounts to a great extent for reduction of exports. Fuel-oil is being used by the railways and also for heating and power purposes on the mainland of British Columbia. According to eminent geologists the time is not far distant when the use of oil for fuel purposes, owing to depletion in oil reserves, will be strongly condemned or prohibited. Notwithstanding this opinion its use is spreading to such an extent, that it is displacing coal under equal conditions. It is a suicidal policy that encourages the importation of fuel-oil into the province of British Columbia, or indeed to any part of western Canada": Robert Strachan, "Coal Mining in British Columbia," *Transactions of the Canadian Institute of Mining and Metallurgy* 26 (1922): 94.

5 Davis, "Forty-Ninth Parallel City," 32.

6 See especially Goodacre, *Dunsmuir's Dream,* 65-68.

7 Davis, "Forty-Ninth Parallel City," 35.

8 Allen Seager and David Roth, "British Columbia and the Mining West: A Ghost of a Chance," in *The Workers' Revolt in Canada, 1917-1925,* ed. Craig Heron (Toronto: University of Toronto Press, 1998), 247.

9 Bowen, *Boss Whistle,* 198.

10 John Norris, "The Vancouver Island Coal Miners, 1912-1914: A Study of an Organizational Strike," *BC Studies* 45 (1980): 71.

11 David W. Sabean, *Power in the Blood: Popular Culture and Village Discourse in Early Modern Germany* (Cambridge: Cambridge University Press, 1984), 29. Sabean's insightful discussion of community is not restricted to early modern Germany but has been applied in other contexts. See also the remarkable book by Inga Clendinnen, *Aztecs: An Interpretation* (Cambridge: Cambridge University Press, 1991), 57-59.

12 David Gilbert writes in the context of British mining disputes: "Women's support groups were the most important innovation in the communal organization of the strike": Gilbert, *Class, Community, and Collective Action* (Oxford: Oxford University Press, 1992), 3.

# *Bibliography*


**Primary Sources**

***British Columbia Archives***

A.F. Buckham Personal and Professional Papers, 1858-1968. MS-0436.

British Columbia. Attorney General. Coroners' Inquiries/Inquests. GR-0431.

British Columbia. Attorney General. Correspondence, 1902-1937. GR-1323.

British Columbia. Attorney General. Correspondence Inward, etc. GR-0429.

British Columbia. Commission on Causes of Explosions in Coal Mines, 1902-1903. GR-0733.

British Columbia. Commission on Claims Arising Out of Riots in 1913 and 1914 on Vancouver Island, 1916. GR-0518.

British Columbia. Commission on Labour, 1912-1914. GR-0684.

British Columbia. County Court (Cumberland). Civil Case Files under the Landlord and Tenant Act. GR-1946.

British Columbia. Provincial Mineralogist. Correspondence Inward. GR-0265.

British Columbia. Provincial Police Force. GR-0068.

British Columbia. Provincial Police Force Daily Reports, 1913-1921. GR-0445.

British Columbia. Provincial Secretary. Correspondence with Municipalities. GR-0566.

Canada. Commission on Unrest and Discontent among Miners and Mine-owners in British Columbia, 1899. GR-1966.

Canadian Collieries (Dunsmuir), Limited: 1860-1906 (MS-0102); 1918-1925 (MS-2175).

Dunsmuir Family Collection. Graham Graham, Collector (Contracts, indentures, bonds, notes and mining licences relating to the Dunsmuir family mining and railroad activities). MS-1012.

John Bryden Fonds. [E/C/B84]. Diary and Letter Book, 1878-80.

John Wood Coburn Papers, 1859-1938 (includes "Early History of the Town of Ladysmith"). MS-0783.

William Wear Johnstone. Records re Wellington Mines, 1910-1958. MS-0780.

***Government Publications***

British Columbia. *Report of the Royal Commission on Labour.* Victoria: King's Printer, 1914.

British Columbia. *Report of the Royal Commission re: Coal in British Columbia.* Victoria: King's Printer, 1914.

British Columbia. *Report of the Special Commission Appointed to Inquire into the Causes of Explosions in Coal Mines.* Sessional Papers, 1903.

British Columbia. "Report on the Wellington Strike," *Journal of the Legislature* (1891).

British Columbia. Sessional Papers. Annual Reports of the Minister of Mines of British Columbia.

British Columbia. *Sessional Papers*. Public Schools Reports.
British Columbia. *Statutes of British Columbia*. 1877.
British Columbia Coal Task Force Technical Committee. *Coal in British Columbia: A Technical Appraisal*. Victoria: The Coal Task Force, 1976.
Canada. *Labour Gazette*.
Canada. *Report of Royal Commissioner on Coal Mining Disputes on Vancouver Island,* Samuel Price, Commissioner. Ottawa: Government Printing Bureau, 1913.
Canada. *Report of the Royal Commission on Chinese and Japanese Immigration*. Ottawa: S.E. Dawson, 1902.
Canada. *Report of the Royal Commission on Chinese Immigration*. Sessional Papers, 1885.
Canada. *Report of the Royal Commission on Industrial Disputes in the Province of British Columbia*. Ottawa: Department of Labour, 1903.
Nova Scotia. *The Westray Story: A Predictable Path to Disaster*. Report of the Westray Mine Public Inquiry, 4 vols., Justice K. Peter Richard, Commissioner. Halifax: Province of Nova Scotia, 1997.

***Nanaimo Community Archives***

Oral History Project.
Western Federation of Miners, Enterprise Union, Minutes, 1903.

*Pamphlets*

Hedley, Rev. John. *The Labor Trouble in Nanaimo District: An Address Given before the Brotherhood of Haliburton Street Methodist Church*. Nanaimo: n.p., 1913.
Kavanagh, Jack. *The Vancouver Island Strike*. Vancouver: BC Miners' Liberation League, 1913.

***Newspapers***

*BC Federationist*
*Ladysmith Chronicle*
*Ladysmith Daily Ledger*
*Ladysmith Leader and Wellington-Extension News*
*Nanaimo Free Press*
*New Westminster News*
*United Mine Workers Journal*
*Vancouver Sun*
*Victoria Daily Colonist*
*Victoria Daily Times*
*Western Clarion*

**Secondary Sources**

Avery, Donald. *"Dangerous Foreigners": European Immigrant Workers and Labour Radicalism in Canada, 1896-1932*. Toronto: University of Toronto Press, 1979.
Baker, William M. "The Miners and the Mounties: The Royal North West Mounted Police and the 1906 Lethbridge Strike." *Labour/Le Travail* 27 (1991): 55-96.
Barman, Jean. *The West Beyond the West*. Toronto: University of Toronto Press, 1991.
Beik, Mildred Allen. *The Miners of Windber: The Struggles of New Immigrants for Unionization, 1890s-1930s*. University Park, PA: Pennsylvania State University Press, 1996.
Belshaw, John Douglas. "British Coalminers on Vancouver Island, 1848-1900: A Social History." PhD thesis, University of London, 1987.

—. "The British Collier in British Columbia: Another Archetype Reconsidered." *Labour/Le Travail* 34 (1994): 11-36.

—. *Colonization and Community: The Vancouver Island Coalfield and the Making of the British Columbian Working Class.* Montreal: McGill-Queen's, 2002.

—. "Cradle to Grave: An Examination of Demographic Behaviour on Two British Columbian Frontiers." *Journal of the Canadian Historical Association* 5 (1994): 41-62.

—. "Mining Technique and Social Division on Vancouver Island, 1848-1900." *British Journal of Canadian Studies* 1 (1986): 45-65.

—. "Provincial Politics, 1871-1916." In *The Pacific Province,* edited by Hugh J.M. Johnston. Vancouver: Douglas and McIntyre, 1996.

—. "The Standard of Living of British Miners on Vancouver Island, 1848-1900." *BC Studies* 84 (1989-90): 37-64.

Benson, John. *British Coalminers in the Nineteenth Century: A Social History.* London: Gill and Macmillan, 1980.

—. "Mining Safety and Miners' Compensation in Great Britain, 1850-1900." In *Sozialgeschichte des Bergbaus im 19. und 20. Jahrhundert,* edited by Klaus Tenfelde. Munich: Beck Verlag, 1992.

Bercuson, David J. *Fools and Wise Men: The Rise and Fall of the One Big Union.* Toronto: McGraw-Hill, 1978.

—. "Labour Radicalism and the Western Industrial Frontier, 1897-1919." *Canadian Historical Review* 58 (1976): 154-75.

Bindocci, Cynthia Gay. "A Comparison of the Roles of Women in Anthracite and Bituminous Mining in Pennsylvania, 1900-1920." In *Sozialgeschichte des Bergbaus im 19. und 20. Jahrhundert,* edited by Klaus Tenfelde. Munich: Beck Verlag, 1992.

Booth, Stephane E. "Ladies in White: Female Activism in the Southern Illinois Coalfields, 1932-1938." In *The United Mine Workers of America: A Model of Industrial Solidarity,* edited by John H.M. Laslett. University Park, PA: Pennsylvania State University Press, 1996.

Bowen, Lynne. *Boss Whistle: The Coal Miners of Vancouver Island Remember.* Lantzville, BC: Oolichan Books, 1982.

—. "Friendly Societies in Nanaimo: The British Tradition of Self-Help in a Canadian Coal-Mining Community." *BC Studies* 118 (1998): 67-92.

—. "The Great Vancouver Island Coal Miners' Strike, 1912-1914." *Journal of the West* 23 (1984): 33-39.

—. "Independent Colliers at Fort Rupert: Labour Unrest on the West Coast, 1849." *The Beaver* 69 (1989): 28-30.

—. *Three Dollar Dreams.* Lantzville, BC: Oolichan Books, 1987.

—. "Towards an Expanded Definition of Oral Testimony: The Coal Miner on Nineteenth-Century Vancouver Island." In *Work, Ethnicity, and Oral History,* edited by Dorothy E. Moore and James H. Morrison. Halifax: International Education Centre, 1988.

Bradbury, Bettina. "Pigs, Cows, and Boarders: Non-Wage Forms of Survival among Montreal Families, 1861-91." *Labour/Le Travail* 14 (1984): 9-48.

Brown, Ivor J. "The Development of Mining Legislation in the United Kingdom – Protection for the Owner, the Miner and the Environment." In *Sozialgeschichte des Bergbaus im 19. und 20. Jahrhundert,* edited by Klaus Tenfelde. Munich: Beck Verlag, 1992.

Brüggemeier, Franz-Josef. *Leben vor Ort: Ruhrbergleute und Ruhrbergbau, 1889-1919.* Munich: Beck Verlag, 1983.

Campbell, Alan B. "Communism and Trade Union Militancy in the Scottish Coalfields." In *Sozialgeschichte des Bergbaus im 19. und 20. Jahrhundert,* edited by Klaus Tenfelde. Munich: Beck Verlag, 1992.

Church, Roy. "Edwardian Labour Unrest and Coalfield Militancy, 1890-1914." *Historical Journal* 30 (1987): 841-57.

—. *The History of the British Coal Industry. Vol. 3, 1830-1913: Victorian Pre-eminence.* Oxford: Clarendon Press, 1986.

Clapp, Charles H. "The Geology of the Nanaimo Coal District." *Transactions of the Canadian Mining Institute* 15 (1912): 334-53.

Colls, Robert. *The Pitmen of the Northern Coalfield: Work, Culture, and Protest, 1790-1850.* Manchester: Manchester University Press, 1987.

Comacchio, Cynthia. "Mechanomorphosis: Science, Management, and 'Human Machinery' in Industrial Canada, 1900-45." *Labour/Le Travail* 41 (1998): 35-67.

Conley, James R. "Frontier Labourers, Crafts in Crisis and the Western Labour Revolt: The Case of Vancouver, 1900-1919." *Labour/Le Travail* 23 (1989): 9-37.

—. "Relations of Production and Collective Action in the Salmon Fishery, 1900-1925." In *Workers, Capital, and the State in British Columbia,* edited by Rennie Warburton and David Coburn. Vancouver: UBC Press, 1988.

Corbin, David Alan. *Life, Work, and Rebellion in the Coal Fields: The Southern West Virginia Miners, 1880-1922.* Chicago: University of Illinois Press, 1981.

Cramer, Michael H. "Public and Political: Documents of the Women's Suffrage Campaign in British Columbia, 1871-1917: The View from Victoria." In *In Her Own Right: Selected Essays on Women's History in British Columbia,* edited by Barbara Latham and Cathy Kess. Victoria: Camosun College, 1980.

Craven, Paul. *"An Impartial Umpire": Industrial Relations and the Canadian State, 1900-1911.* Toronto: University of Toronto Press, 1980.

Creese, Gillian. "Class, Ethnicity, and Conflict: The Case of Chinese and Japanese Immigrants, 1880-1923." In *Workers, Capital, and the State in British Columbia,* edited by Rennie Warburton and David Coburn. Vancouver: UBC Press, 1988.

Cross, Michael A., and Gregory S. Kealey, eds. *The Consolidation of Capitalism, 1896-1929.* Toronto: McClelland and Stewart, 1983.

Davis, Isabelle. "Forty-Ninth Parallel City: An Economic History of Ladysmith." BA thesis, University of British Columbia, 1953.

Diamond, Sara. "A Union Man's Wife: The Ladies' Auxiliary Movement of the IWA, the Lake Cowichan Experience." In *Not Just Pin Money,* edited by Barbara K. Latham and Roberta J. Pazdro. Victoria: Camosun College, 1984.

Dick, W.J. *Conservation of Coal in Canada.* Toronto: Bryant Press, 1914.

—. *Mine-Rescue Work in Canada.* Ottawa: Commission of Conservation, 1912.

Dix, Keith. "Mechanization, Workplace Control, and the End of the Hand-Loading Era." In *The United Mine Workers of America: A Model of Industrial Solidarity,* edited by John H.M. Laslett. University Park, PA: Pennsylvania State University Press, 1996.

—. *What's a Coal Miner to Do? The Mechanization of Coal Mining.* Pittsburgh: University of Pittsburgh Press, 1988.

—. *Work Relations in the Coal Industry: The Hand-Loading Era, 1880-1930.* Morgantown, WV: Institute for Labor Studies, 1977.

Engels, Friedrich. *The Condition of the Working Class in England.* London: Penguin Books, 1987.

Fishback, Price V. "Did Miners 'Owe Their Souls to the Company Store'?" *Journal of Economic History* 46 (1986): 1011-29.

—. "The Economics of Company Housing." *Journal of Law, Economics, and Organization* 8 (1992): 346-65.

—. "The Miner's Work Environment: Safety and Company Towns in the Early 1900s." In *The United Mine Workers of America: A Model of Industrial Solidarity,* edited by John H.M. Laslett. University Park, PA: Pennsylvania State University Press, 1996.

—. *Soft Coal, Hard Choices: The Economic Welfare of Bituminous Coal Miners, 1890-1930.* New York: Oxford University Press, 1993.

—. "Workplace Safety during the Progressive Era: Fatal Accidents in Bituminous Coal Mining, 1912-23." *Explorations in Economic History* 23 (1986): 269-98.

Fleming, R.B. *The Railway King of Canada: Sir William Mackenzie, 1849-1923.* Vancouver: UBC Press, 1991.

Foner, Philip S., ed. *Mother Jones Speaks: Collected Writings and Speeches.* New York: Monad Press, 1983.

Forestell, Nancy M. "The Miner's Wife: Working-Class Femininity in a Masculine Context, 1920-1950." In *Gendered Pasts: Historical Essays in Femininity and Masculinity in Canada,* edited by Kathryn McPherson, Cecilia Morgan, and Nancy M. Forestell. Toronto: Oxford University Press, 1999.

Forsey, Eugene. *Trade Unions in Canada, 1812-1902.* Toronto: University of Toronto Press, 1982.

Fox, Maier B. *United We Stand: The United Mine Workers of America, 1890-1990.* Washington, DC: United Mine Workers of America, 1990.

Frank, David. "Contested Terrain: Workers' Control in the Cape Breton Coal Mines in the 1920s." In *On the Job: Confronting the Labour Process in Canada,* edited by Craig Heron and Robert Storey. Montreal and Kingston: McGill-Queen's University Press, 1986.

—. "Industrial Democracy and Industrial Legality: The UMWA in Nova Scotia, 1908-1927." In *The United Mine Workers of America: A Model of Industrial Solidarity,* edited by John H.M. Laslett. University Park, PA: Pennsylvania State University Press, 1996.

—. "The Miners' Financier: Women in the Cape Breton Coal Towns, 1917." *Atlantis* 8 (1983): 137-43.

Gallacher, Daniel T. "Men, Money, Machines: Studies Comparing Colliery Operations and Factors of Production in British Columbia's Coal Industry to 1891." PhD thesis, University of British Columbia, 1979.

Geary, Dick. *European Labour Protest, 1848-1939.* London: Methuen, 1981.

—. "Protest and Strike: Recent Research on 'Collective Action' in England, Germany, and France." In *Arbeiter und Arbeiterbewegung im Vergleich,* edited by Klaus Tenfelde. Munich: Beck Verlag, 1986.

Gilbert, David. *Class, Community, and Collective Action.* Oxford: Oxford University Press, 1992.

Goodacre, Richard. *Dunsmuir's Dream: Ladysmith, The First Fifty Years.* Victoria: Porcépic Books, 1991.

Grove, Alan, and Ross Lambertson. "Pawns of the Powerful: The Politics of Litigation in the Union Colliery Case." *BC Studies,* 103 (1994): 3-31.

Haimson, Leopold H., and Charles Tilly, eds. *Strikes, Wars, and Revolutions in an International Perspective.* Cambridge: Cambridge University Press, 1989.

Hak, Gordon. "The Socialist and Labourist Impulse in Small-Town British Columbia: Port Alberni and Prince George, 1911-33." *Canadian Historical Review* 70 (1989): 519-42.

Hardesty, Donald L. "The Miner's Domestic Household: Perspectives from the American West." In *Sozialgeschichte des Bergbaus im 19. und 20. Jahrhundert,* edited by Klaus Tenfelde. Munich: Beck Verlag, 1992.

Harrison, Royden, ed. *Independent Collier: The Coal Miner as Archetypical Proletarian Reconsidered.* London: Harvester Press, 1978.

Hartewig, Karin. *Das unberechenbare Jahrzehnt: Bergarbeiter und ihre Familien im Ruhrgebiet, 1914-1924.* Munich: Beck Verlag, 1993.

Haynes, M.J. "Strikes." In *The Working Class of England, 1875-1914,* edited by John Benson. London: Croom Helm, 1985.

Heron, Craig. "Labourism and the Canadian Working Class." *Labour/Le Travail* 13 (1984): 45-75.

Hickey, Stephen. "Bergmannsarbeit an der Ruhr vor dem Ersten Weltkrieg." In *Glück auf, Kamaraden! Die Bergarbeiter und ihrer Organisation in Deutschland,* edited by Hans Mommsen and Ulrich Borsdorf. Cologne: Bund-Verlag, 1979.

—. "The Shaping of the German Labour Movement: Miners in the Ruhr." In *Society and Politics in Wilhelmine Germany,* edited by Richard J. Evans. London: Croom Helm, 1978.

—. *Workers in Imperial Germany: The Miners of the Ruhr.* Oxford: Clarendon Press, 1985.

Hinde, John R. "'Stout Ladies and Amazons': Women in the British Columbia Coal-Mining Community of Ladysmith, 1912-14." *BC Studies* 114 (1997): 33-57.

Howard, Irene. *The Struggle for Social Justice in British Columbia: Helena Gutteridge, the Unknown Reformer.* Vancouver: UBC Press, 1992.

Jameson, Elizabeth. *All that Glitters: Class, Conflict, and Community in Cripple Creek.* Urbana: University of Illinois Press, 1998.

—. "Imperfect Unions: Class and Gender in Cripple Creek, 1894-1904." In *Class, Sex, and the Woman Worker,* edited by Milton Cantor and Bruce Laurie. Westport, CT: Greenwood Press, 1977.

Jamieson, Stuart M. *Times of Trouble: Labour Unrest and Industrial Conflict in Canada, 1900-66.* Ottawa: Task Force on Labour Relations, 1968.

John, Angela. *By the Sweat of Their Brow: Women Workers in Victorian Coal Mines.* London: Croom Helm, 1980.

Johnson-Cull, Viola, comp. *Chronicle of Ladysmith and District.* [Ladysmith]: Ladysmith New Horizons Historical Society, 1980.

Johnstone, Bill. *Coal Dust in My Blood: The Autobiography of a Coal Miner.* Victoria: Royal British Columbia Museum, 1993.

Juricic, Zelimir B. "Croatians Enlivened Mining Towns." *BC Historical News* 27 (Summer 1994): 22-25.

—. "Croatians Killed in Ladysmith Mine Blast." *BC Historical News* 26 (Winter 1992-93): 20-23.

Kealey, Gregory S. "1919: The Canadian Labour Revolt." *Labour/Le Travail* 13 (1984): 11-44.

Kealey, Linda. "Canadian Socialism and the Woman Question, 1900-1914." *Labour/Le Travail* 13 (1984): 77-100.

—. *Enlisting Women for the Cause: Women, Labour, and the Left in Canada, 1890-1920.* Toronto: University of Toronto Press, 1998.

Keelor, John Thomas. "The Price of Lives and Limbs Lost at Work: The Development of No-Fault Workers' Compensation Legislation in British Columbia, 1910 to 1916." MA thesis, University of Victoria, 1996.

Kerr, Clark, and Abraham Siegal. "The Interindustry Propensity to Strike: An International Comparison." In *Industrial Conflict,* edited by Arthur Kornhauser, Robert Dubin, and Arthur M. Ross. New York: McGraw-Hill, 1954.

Lamb, H. Mortimer. "The Coal Industry of Vancouver Island, British Columbia." *British Columbia Mining Record* 4 (1898): 14-19, 29-32.

Lambertson, Ross. "After Union Colliery: Law, Race, and Class in the Coalmines of British Columbia." In *Essays in the History of Canadian Law,* vol. 6, edited by Hamar Foster and John McLaren. Toronto: University of Toronto Press, 1995.

Laslett, John H.M. "British Immigrant Colliers, and the Origins and Early Development of the UMWA." In *The United Mine Workers of America: A Model of Industrial Solidarity,* edited by John H.M. Laslett. University Park, PA: Pennsylvania State University Press, 1996.

—, ed. *The United Mine Workers of America: A Model of Industrial Solidarity.* University Park, PA: Pennsylvania State University Press, 1996.

Leier, Mark. "Ethnicity, Urbanism, and the Labour Aristocracy: Rethinking Vancouver Trade Unionism, 1889-1909." *Canadian Historical Review* 74 (1993): 510-34.

—. *Rebel Life: The Life and Times of Robert Gosden, Revolutionary, Mystic, Labour Spy.* Vancouver: New Star Books, 1999.

—. *Red Flags and Red Tape: The Making of a Labour Bureaucracy.* Toronto: University of Toronto Press, 1995.

—. *Where the Fraser River Flows: The Industrial Workers of the World in British Columbia.* Vancouver: New Star Books, 1990.

—. "W[h]ither Labour History: Regionalism, Class, and the Writing of BC History." *BC Studies* 111 (1996): 61-75.

Lewis, Brian. *Coal Mining in the Eighteenth and Nineteenth Centuries.* London: Longman, 1971.

Lewis, Ronald L. *Black Coal Miners in America: Race, Class and Community Conflict.* Lexington, KY: University Press of Kentucky, 1987.

Long, Priscilla. "The 1913-1914 Colorado Fuel and Iron Strike, with Reflections on the Causes of Coal-Strike Violence." In *The United Mine Workers of America: A Model of Industrial Solidarity,* edited by John H.M. Laslett. University Park, PA: Pennsylvania State University Press, 1996.

—. *Where the Sun Never Shines: A History of America's Bloody Coal Industry.* New York: Paragon House, 1989.

—. "The Women of the Colorado Fuel and Iron Strike, 1913-14." In *Women, Work and Protest: A Century of US Women's Labor History,* edited by Ruth Milkman. Boston: Routledge and Kegan Paul, 1985.

Loo, Tina. *Making Law, Order, and Authority in British Columbia, 1821-1871.* Toronto: University of Toronto Press, 1994.

Loosmore, Thomas Robert. "The British Columbia Labor Movement and Political Action, 1879-1906." MA thesis, University of British Columbia, 1954.

Lutz, John. "After the Fur Trade: The Aboriginal Labouring Class of British Columbia, 1849-1890." *Journal of the Canadian Historical Association* 2 (1992): 69-94.

—. "Losing Steam: The Boiler and Engine Industry as an Index of British Columbia's Deindustrialization, 1880-1915." Canadian Historical Association *Historical Papers* (1988): 182-202.

Mackie, Richard Somerset. *The Wilderness Profound: Victorian Life on the Gulf of Georgia.* Victoria: Sono Nis Press, 1995.

MacLeod, Donald. "Colliers, Colliery Safety and Coal Mine Politics: The Nova Scotian Experience, 1873 to 1910." Canadian Historical Association *Historical Papers* (1983): 226-53.

Malcolmson, John. "Politics and the State in the Nineteenth Century." In *Workers, Capital, and the State in British Columbia,* edited by Rennie Warburton and David Coburn. Vancouver: UBC Press, 1988.

McCormack, A. Ross. "The Emergence of the Socialist Movement in British Columbia." *BC Studies* 21 (1974): 3-27.

—. "The Industrial Workers of the World in Western Canada." Canadian Historical Association *Historical Papers* (1975): 173-84.

—. *Reformers, Rebels, and Revolutionaries: The Western Canadian Radical Movement, 1899-1919.* Toronto: University of Toronto Press, 1977.

McDonald, Robert A.J. *Making Vancouver, 1863-1913.* Vancouver: UBC Press, 1996.

—. "Victoria, Vancouver, and the Economic Development of British Columbia, 1886-1914." In *British Columbia Historical Readings,* edited by W. Peter Ward and Robert A.J. McDonald. Vancouver: Douglas and McIntyre, 1981.

—. "Working Class Vancouver, 1886-1914: Urbanism and Class in British Columbia." *BC Studies* 69-70 (1986): 33-69.

McIntosh, Robert. "The Boys in the Nova Scotian Coal Mines, 1873-1923." *Acadiensis* 16 (1987): 35-50.

—. *Boys in the Pits: Child Labour in Coal Mines.* Montreal and Kingston: McGill-Queen's University Press, 2000.

—. "The Family Economy and Boy Labour, Sydney Mines, Nova Scotia, 1871-1901." *Nova Scotia Historical Review* 13 (1993): 87-100.

McKay, Ian. "'By Wisdom, Wile, or War': The Provincial Workmen's Association and the Struggle for Working-Class Independence in Nova Scotia, 1879-1897." *Labour/Le Travail* 18 (1986): 13-62.

—. "Coal Miners and the Longue Durée: Learning from Decazeville." *Labour/Le Travail* 20 (1987): 223.

—. "The Realm of Uncertainty: The Experience of Work in the Cumberland Coal Mines, 1873-1927." *Acadiensis* 16 (1986): 3-57.

—. "Strikes in the Maritimes, 1901-1914." *Acadiensis* 13 (1983): 3-46.

McPhail, Clark. *The Myth of the Madding Crowd.* New York: A. de Gruyter, 1991.

Michel, Joel. "Bergarbeiter-Kommunen und Patriarchalismus in Westeuropa vor 1914." In *Sozialgeschichte des Bergbaus im 19. und 20. Jahrhundert,* edited by Klaus Tenfelde. Munich: Beck Verlag, 1992.

Mommsen, Hans. "Soziale und politische Konflikte an der Ruhr 1905 bis 1924." In *Arbeiterbewegung und industrieller Wandel,* edited by Hans Mommsen. Wuppertal, Germany: Peter Hammer Verlag, 1980.

Mouat, Jeremy. "The Genesis of Western Exceptionalism: British Columbia's Hard-Rock Miners, 1985-1903." *Canadian Historical Review* 71 (1990): 317-45.

—. "The Politics of Coal: A Study of the Wellington Miners' Strike of 1890-91." *BC Studies* 77 (1988): 3-29.

—. *Roaring Days: Rossland's Mines and the History of British Columbia.* Vancouver: UBC Press, 1995.

Muise, Delphin A., and Robert G. McIntosh. *Coal Mining in Canada: A Historical and Comparative Overview.* Ottawa: National Museum of Science and Technology, 1996.

Nelles, H.V. "Horses of a Shared Colour: Interpreting Class and Identity in Turn-of-the-Century Vancouver." *Labour/Le Travail* 40 (1997): 269-75.

Neville, Robert G. "The Courrières Colliery Disaster, 1906." *Journal of Contemporary History* 13 (1978): 33-52.

Norris, John. "The Vancouver Island Coal Miners, 1912-1914: A Study of an Organizational Strike." *BC Studies* 45 (1980): 56-72.

Ormsby, Margaret A. *British Columbia: A History.* Toronto: Macmillan, 1958.

Palmer, Bryan D. *Working Class Experience: Rethinking the History of Canadian Labour, 1800-1991.* Toronto: McClelland and Stewart, 1992.

Palmer, Bryan D., and Craig Heron. "Through the Prism of Strike: Industrial Conflict in Southern Ontario, 1901-1914." *Canadian Historical Review* 58 (1977): 423-58.

Paterson, D.G. "European Finance Capital and British Columbia: An Essay on the Role of the Regional Entrepreneur." In *British Columbia Historical Readings,* edited by W. Peter Ward and Robert A.J. McDonald. Vancouver: Douglas and McIntyre, 1981.

Penfold, Steven. "'Have You No Manhood in You?': Gender and Class in the Cape Breton Coal Towns, 1920-1926." *Acadiensis* 23 (1994): 21-44.

Pentland, H. Clare. "The Western Canadian Labour Movement, 1897-1919." *Canadian Journal of Political and Social Theory* 3 (1979): 53-78.

Phillips, Paul. *No Power Greater: A Century of Labour in British Columbia.* Vancouver: Boag Foundation, 1967.

—. "The Underground Economy: The Mining Frontier to 1920." In *Workers, Capital, and the State in British Columbia: Selected Papers,* edited by Rennie Warburton and David Coburn. Vancouver: UBC Press, 1988.

Planta, S.P. "The Coal Fields of Vancouver Island, British Columbia." *Canadian Mining Manual* (1893): 279-97.

Porsild, Charlene. *Gamblers and Dreamers: Women, Men, and Community in the Klondike.* Vancouver: UBC Press, 1998.

Poulantzas, Nicos. *Political Power and Social Classes.* London: Sheed and Ward, 1973.

Ralston, H. Keith. "Miners and Managers: The Organization of Coal Production on Vancouver Island by the Hudson's Bay Company, 1848-1862." In *The Company on the Coast,* edited by E. Blanche Norcross. Nanaimo: Nanaimo Historical Society, 1983.

Ramsey, Bruce. *The Noble Cause: The Story of the United Mine Workers of America in Western Canada.* Calgary: UMWA District 18, 1990.

Reid, David J. "Company Mergers in the Fraser River Salmon Canning Industry." In *British Columbia Historical Readings,* edited by W. Peter Ward and Robert A.J. McDonald. Vancouver: Douglas and McIntyre, 1981.

Reid, Donald. *The Miners of Decazeville.* Cambridge, MA: Harvard University Press, 1985.

—. "The Role of Mine Safety in the Development of Working-Class Consciousness and Organization: The Case of the Aubin Coal Basin, 1867-1914." *French Historical Studies* 13 (1981): 98-119.

Robin, Martin. *The Rush for Spoils: The Company Province, 1871-1933.* Toronto: McClelland and Stewart, 1972.

Roy, Patricia E. "Progress, Prosperity and Politics: The Railway Policies of Richard McBride." *BC Studies* 47 (1980): 3-28.

—. *A White Man's Province: British Columbia Politicians and Chinese and Japanese Immigrants, 1858-1914.* Vancouver: UBC Press, 1989.

Rutherdale, Robert. "Canada's August Festival: Communitas, Liminality, and Social Memory." *Canadian Historical Review* 77 (1996): 221-49.

Sangster, Joan. *Dreams of Equality: Women on the Canadian Left, 1920-1950.* Toronto: McClelland and Stewart, 1989.

Schofield, Ann. "An 'Army of Amazons': The Language of Protest in a Kansas Mining Community, 1921-22." *American Quarterly* 37 (1985): 686-701.

Schorske, Carl E. *German Social Democracy, 1905-1917: The Development of the Great Schism.* Cambridge, MA: Harvard University Press, 1955.

Schwantes, Carlos. *Radical Heritage, Labor, Socialism and Reform in Washington and British Columbia, 1885-1917.* Seattle: University of Washington Press, 1979.

Schwieder, Dorothy. *Black Diamonds: Life and Work in Iowa's Coal Mining Communities, 1895-1925.* Ames: Iowa State University Press, 1983.

Seager, Allen. "Miners' Struggles in Western Canada: Class, Community, and the Labour Movement, 1890-1930." In *Class, Community and the Labour Movement: Wales and Canada, 1850-1930,* edited by Deian R. Hopkin and Gregory S. Kealey. Oxford: Oxford University Press, 1989.

—. "The Resource Economy, 1871-1921." In *The Pacific Province,* edited by Hugh J.M. Johnston. Vancouver: Douglas and McIntyre, 1996.

—. "Socialists and Workers: The Western Canadian Coal Miners, 1900-1921." *Labour/Le Travail* 16 (1985): 23-59.

Seager, Allen, and Adele Perry. "Mining the Connections: Class, Ethnicity, and Gender in Nanaimo, British Columbia, 1891." *Histoire sociale/Social History* 30 (1997): 55-76.

Sharp, Alexander. "Some Notes on the Longwall Method of Mining Coal." *Transactions of the Canadian Mining Institute* 16 (1913): 463-501.

Shubert, Adrian. *The Road to Revolution in Spain: The Coal Miners of Asturias, 1860-1934.* Chicago: University of Illinois Press, 1987.

*Souvenir of the 50th Anniversary of the Corporation of the City of Ladysmith, 1904-54.* Ladysmith: Ladysmith Chamber of Commerce, 1954.

Strachan, Robert. "Coal Mining in British Columbia." *Transactions of the Canadian Institute of Mining and Metallurgy* 26 (1922): 70-132.

Tenfelde, Klaus. "The Herne Riots of 1899." In *The Social History of Politics: Critical Perspectives in West German Historical Writing since 1945,* edited by Georg G. Iggers. Leamington Spa, UK: Berg Publishers, 1985.

—. *Sozialgeschichte der Bergarbeiterschaft an der Ruhr im 19. Jahrhundert.* Bonn-Bad Godesberg: Verlag Neue Gesellschaft, 1977.

—, ed. *Sozialgeschichte des Bergbaus im 19. und 20. Jahrhundert.* Munich: Beck Verlag, 1992.

Thompson, E.P. "The Moral Economy of the English Crowd." In *Customs in Common.* New York: New Press, 1993.

Trempé, Rolande. *Les mineurs de Carmaux: 1848-1914,* 2 vols. Paris: Les éditions ouvrières, 1971.

Trotter, Joe W. *Coal, Class, and Color: Blacks in Southern West Virginia, 1905-1932.* Urbana: University of Illinois Press, 1990.

Tuck, J.H. "The United Brotherhood of Railway Employees in Western Canada, 1898-1905." *Labour/Le Travail* 11 (1983): 63-88.

Tucker, Eric. *Administering Danger in the Workplace: The Law and Politics of Occupational Health and Safety Regulation in Ontario, 1850-1914.* Toronto: University of Toronto Press, 1990.

Volkmann, Heinrich, and Jürgen Bergmann, eds. *Sozialer Protest: Studien zu traditioneller Resistenz und kollektiver Gewalt in Deutschland.* Opladen: Westdeutscher Verlag, 1984.

Warburton, Rennie. "The Workingmen's Protective Association, Victoria, BC, 1878: Racism, Intersectionality and Status Politics." *Labour/Le Travail* 43 (1999): 105-20.

Warburton, Rennie, and David Coburn, eds. *Workers, Capital, and the State in British Columbia: Selected Papers.* Vancouver: UBC Press, 1988.

Ward, Peter. "Race and Class in the Social Structure of British Columbia, 1870-1914." *BC Studies* 45 (1980): 17-35.

—. *White Canada Forever: Popular Attitudes and Public Opinion towards Orientals in British Columbia.* Montreal and Kingston: McGill-Queen's University Press, 1978.

Wargo, Alan. "The Great Coal Strike: The Vancouver Island Coal Miners' Strike, 1912-14." BA thesis, University of British Columbia, 1962.

Williams, David Ricardo. *Call in Pinkerton's: American Detectives at Work for Canada.* Toronto: Dundurn Press, 1998.

Yarmie, Andrew. "The Right to Manage: Vancouver Employers' Associations, 1900-1923." *BC Studies* 90 (1991): 40-74.

—. "The State and Employers' Associations in British Columbia: 1900-1932." *Labour/Le Travail* 45 (2000): 53-101.

# Index

Set in New Baskerville and Galliard by Artegraphica Design Co.

Text design: Irma Rodriguez

Printed and bound in Canada by Friesens

Copy editor: Frank Chow

Proofreader: Deborah Kerr

Cartographer: Eric Leinberger